ANALOG INTEGRATED CIRCUITS AND SIGNAL PROCESSING

An International Journal

Volume 17—1998

SPRINGER SCIENCE+BUSINESS MEDIA, LLC 1998

ANALOG INTEGRATED CIRCUITS AND SIGNAL PROCESSING

An International Journal

Volume 17, No 1/2, September 1998

Special Issue: Field Programmable Analog Arrays
Guest Editors: Edmund Pierzchala, Glenn Gulak, Leon O. Chua and Angel Rodríguez-Vázquez

Library of Congress Cataloging-in-Publication Data

A C.I.P. Catalogue record for this book is available
from the Library of Congress.

DOI 10.1007/978-1-4757-5224-3

Printed on acid-free paper.
www.springer.com/mycopy

Analog Integrated Circuits and Signal Processing, 17, 5–6 (1998)

Editorial

This special issue presents an emerging technology of field-programmable analog arrays, FPAAs. The term was coined by Lee and Gulak, and it is reminiscent of the field-programmable gate arrays, FPGAs. While FPGAs are a mature technology, with an annual market share of over a billion dollars, FPAAs are still a solution looking for a problem. Given that FPGAs were not widely accepted until early 1990s, over five years after Xilinx introduced their first FPGA product into the market, it may be some time before FPAAs make big headlines. Before this new technology can be widely accepted, it should prove its maturity, and its superiority over existing solutions.

In a way, FPAAs are nothing new: they are just another attempt to provide flexibility to analog circuits, much like analog computers, which were seemingly abandoned decades ago.

Analog computers could not evolve the same way digital computers did, i.e. by mixing program and data in the Von-Neumann paradigm. Today, digital FPGAs introduce a new dimension to programming, namely changing the function of a system by means of modifying its configuration. This new programming paradigm is taken up by the FPAAs.

With artificial neural networks (ANNs), cellular neural networks (CNNs), silicon retinas and cochleas leading the way in modern analog processors, FPAAs are placing themselves somewhere in between smart analog circuits, a kind of super-op amps good for "all" analog designs, and universal analog signal processors. As such, FPAAs may become a medium for the implementation of ANNs, CNNs, and other computing paradigms, those mimicking biological systems, and those conceived entirely by the human mind much like a modern microprocessor is a medium for implementation of digital algorithms.

The power and the weaknesses of the FPAAs stem from the same source: their use of analog representation of information. The infinite abundance of analog states, the astonishing simplicity of the realization of some important signal-processing operations, such as integration, multiplication, addition, and great speed, can only be envied by digital processors, which will keep pushing the limits of technology toward higher speeds and smaller circuit sizes, only to give the *analog* circuits even greater speed and size advantage. Naturally, there is a price to be paid by FPAAs.

This issue starts with an excellent review of FPAA design approaches, both academic and industrial, by D'Mello and Gulak.

The next four papers describe switched-capacitor (SC) implementations of FPAAs with various architectures: The first two in this group, by Lee and Hui, and Kutuk and Kang, present two different approaches resulting from academic research, while the two others, by Bratt and Macbeth, and Klein, describe industrial designs.

CMOS designs by Premont et al. and Embabi et al. open the last group of four papers which describe continuous-time (c-t) FPAAs, all from academia.

Finally, the two papers by Pierzchala and Perkowski present the design and applications of a high-frequency FPAA based on a BJT technology.

Edmund Pierzchala
Glenn Gulak
Leon O. Chua
Angel Rodríguez-Vázquez

Edmund Pierzchala received his M.S. degree in electronic engineering from Warsaw University of Technology, Warsaw, Poland. He worked as a research assistant and a senior research assistant in the Institute of Biocybernetics and Biomedical Engineering of Polish Academy of Sciences in Warsaw, Poland, and the Nuclear Research Institute in Świerk, Poland, in the areas of knowledge-based systems and pattern recognition. Presently, he is completing a Ph.D. degree at the Department of Electrical Engineering of Portland State University, where he taught a number of undergraduate and graduate courses in EE. His research interests include programmable analog circuits, design automation, analog and mixed-signal circuits design, modeling, and simulation.

He consulted for Cypress Semiconductor Corp. as a member of the development team of WARP, the first VHDL compiler for EPLDs. He is a co-founder of Analogix Corp., an R&D startup company formed to develop and commercialize high-speed FPAAs. Presently, he is working for the Modeling Department of Analogy, Inc., where he develops simulation models and characterization procedures and software for integrated circuits, components and devices.

Analog Integrated Circuits and Signal Processing, 17, 7–34 (1998)

Design Approaches to Field-Programmable Analog Integrated Circuits

DEAN R. D'MELLO AND P. GLENN GULAK

Department of Electrical and Computer Engineering, University of Toronto, Toronto, Ontario Canada M5S 3G4

Received July 15, 1996; Accepted June 11, 1997

Abstract. The drive towards shorter design cycles for analog integrated circuits has given impetus to several developments in the area of Field-Programmable Analog Arrays (FPAAs). Various approaches have been taken in implementing structural and parametric programmability of analog circuits. Recent extensions of this work have married FPAAs to their digital counterparts (FPGAs) along with data conversion interfaces, to form Field-Programmable Mixed-Signal Arrays (FPMAs). This survey paper reviews work to date in the area of programmable analog and mixed-signal circuits. The body of work reviewed includes university and industrial research, commercial products and patents. A time-line of important achievements in the area is drawn, the status of various activities is summarized, and some directions for future research are suggested.

Key Words: FPAA, FPMA, FPGA, programmable analog, IC, analog CAD, field-programmable, mixed-signal IC

1. Introduction

The role of analog integrated circuits in modern electronic systems remains important, even though digital circuits dominate the market for VLSI solutions. Analog systems have always played an essential role in interfacing digital electronics to the real world in applications such as analog signal processing and conditioning, industrial process and motion control and biomedical measurements. In addition, analog solutions are becoming increasingly competitive with digital circuits for dense, low-power, high-speed applications in low-precision signal-processing. An important advantage of digital integrated circuits has been their relative ease of design over analog circuits. In particular, since digital circuit design is amenable to automation, several CAD-compatible digital integrated circuit design methodologies have been developed, including design-for-testability, design optimization, rapid prototyping in Field-Programmable Gate Arrays (FPGAs) and, more recently, hardware synthesis from behavioral descriptions. In the highly competitive electronics industry, the application of CAD techniques to digital integrated circuit design has led to shorter design cycles, alleviating some of the time-to-market pressures felt by developers of commercial products.

Because of the wide variety of analog functions required in electronic systems and the complexity of the signals (frequency, time, signal levels, parasitics), analog system design is very specialized and supported by a diverse set of CAD tools that are more difficult to integrate than those required for digital design. The drive towards shorter design cycles for analog integrated circuits has demanded the development of high performance analog circuits that are reconfigurable and suitable for CAD methodologies [1].

This has been the motivation for research in the area of Field-Programmable Analog Arrays (FPAAs), which seek to provide accurate, low-cost, rapid-prototyping techniques for analog and mixed analog-digital circuits—a long awaited development for circuit designers. Commercial products introduced recently, along with progress made at University research labs, indicate renewed interest and further accomplishment in achieving this goal. This paper reviews work to date in the area of Field-Programmable Analog Arrays and Field-Programmable Mixed-Signal Arrays (FPMAs). We begin with a general description of FPAAs, followed by a discussion of some of the architecture, circuit design, and implementation issues of programmability in analog circuits.

1.1. General Description of FPAAs

In its most general form, an FPAA is a monolithic collection of analog building blocks, a user-controllable routing network used for passing signals between the building blocks and a collection of memory elements used to define both the function and structure. Alternatively, the structure may be defined by other means such as antifuse programming. Fig. 1 shows a conceptual block diagram of an FPAA, including a set of Configurable Analog Blocks (CABs) and a routing net-work. Configuration memory is provided for the blocks and interconnect. Associated with this is a CAD system, as shown in Fig. 2, that takes the designer's circuit and translates it into a collection of configuration bits that, when stored in the memory elements, instantiates the circuit in the programmable array. Several approaches to realizing such a system have been attempted.

The next step in the evolution of field-programmable analog systems is undoubtedly the integration of analog and digital functions on a single chip, to create a Field-Programmable Mixed-Signal Array (FPMA). One of the characteristics a designer would like in such a concept is to freely exchange analog and digital signals within the prototyping medium. Market research firms have forecast that mixed-signal ICs will represent 30% of the $12.7 billion standard-cell IC market in the year 2000, up from 26% of the $4.8 billion market in 1995 [2]. This indicates that there might be a promising market for a rapid-prototyping medium for mixed-signal ICs.

In this paper we present a detailed survey of work to date in the area of programmable analog and mixed-signal integrated circuits, with a view to introducing the reader to the field of Field-Programmable Analog and Mixed-Signal Arrays. The body of work reviewed includes approaches to making analog circuits programmable both in structure (to implement different circuit topologies) and parameters (to implement variable component values, amplifier gains, etc.). It should be noted, however, that the field of analog circuits with programmable parameters includes numerous publications in areas such as Automatic Gain Control (AGC), programmable amplifiers and filters, from which just a few examples are addressed in this survey of university and industrial research, commercial products and patents.

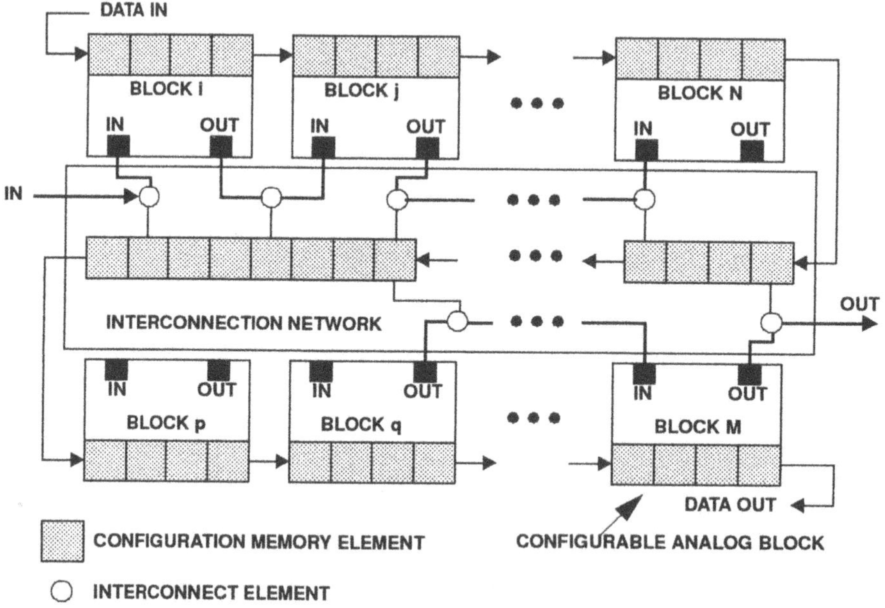

Fig. 1. FPAA Conceptual Block Diagram.

1.2. FPAA Design Issues

In this section, programmable analog integrated circuit design issues are discussed to provide a foundation on which to base the descriptions of research and commercial work that follow.

1.2.1. Discrete-time vs. Continuous-time. A key choice in the implementation of an FPAA is whether to operate in discrete-time or continuous-time.

Discrete-time approaches, such as switched-capacitor circuit techniques, are well suited to digital control and hence do not require the use of on-chip tuning circuitry for VLSI implementations of programmable components. However, such sampled-data techniques require that input signals be bandlimited to at least one half of the sampling frequency, and hence anti-aliasing and reconstruction filters must be used. This requirement often limits the bandwidth of discrete-time FPAA circuit implementations. The integrated circuits described in the literature that operate in discrete-time employ various techniques, namely: switched-capacitor circuits [3,4], controlled duty-cycle signal chopping and reconstruction [5], analog to digital conversion followed by digital processing and digital to analog conversion [6], or switched-current circuits [7].

Continuous-time circuit techniques [8–12] do not need bandlimited input signals, but may require more complicated implementations to have circuit components programmable over a large dynamic range. Continuous-time techniques of both subthreshold and linear circuits have been used in programmable analog circuits.

1.2.2. Voltage-mode vs. Current-mode. Another important design choice is whether to use voltage or current as the signal parameter in the FPAA implementation. Voltage signals have a high fanout, and voltage-mode circuit techniques are well-developed. Several programmable analog circuits have been developed based on voltage mode signals [3–5,8,9,13].

However, advantages such as the simpler implementations of current mode circuit operations (e.g. algebraic addition can be performed simply by wiring signals together), and the high accuracy and high bandwidth of current-mode amplifier circuits [14], have led to the choice of current as a signal parameter for some implementations of programmable analog

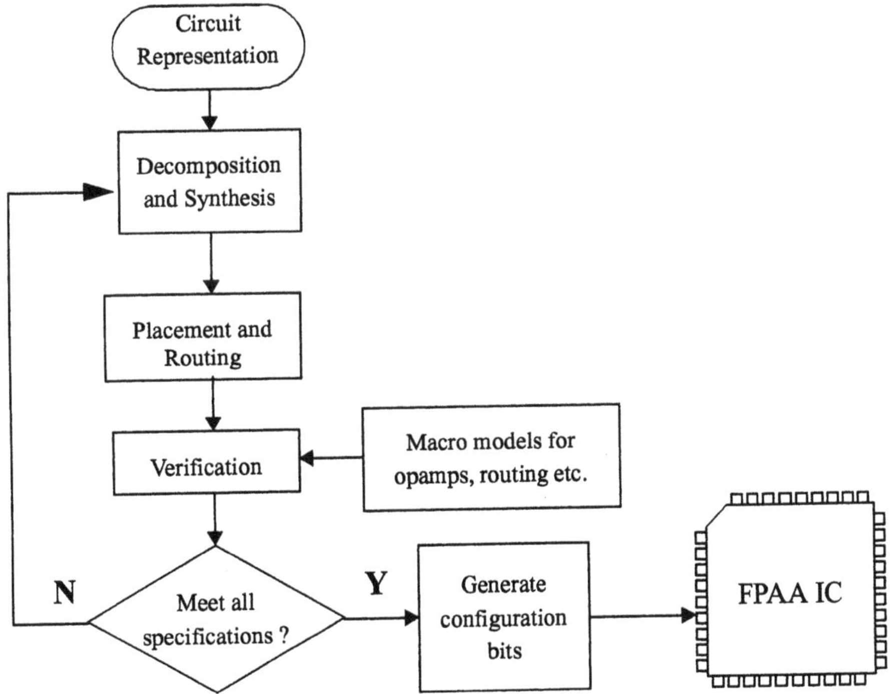

Fig. 2. The FPAA Design Process (from [18]).

circuits [7,10–12]. Recent trends towards lower power supply voltages have reduced the dynamic range available in voltage-mode circuits, making current-mode signalling more attractive [15]. In this context, the concept of "adjoint networks" [15] is interesting in that a transformation between voltage- and current-mode implementations of circuits is possible, and an adjoint realization of a circuit may lead to superior noise performance. Mixed voltage- and current-mode approaches have also been attempted [16].

While both discrete- and continuous-time methods are possible for voltage- and current-modes, discrete-time approaches have been predominantly voltage-mode. Discrete-time current-mode approaches such as switched-current circuits [17] might provide approaches for future programmable analog circuits, as demonstrated in [7].

1.2.3. CAB Design. The design of the Configurable Analog Block (CAB), the basic cell used in FPAAs, is usually influenced by a number of factors, including the functionality and performance features of circuits to be prototyped, the area-efficiency of routing resources dictated by the CAB design itself and the supporting semiconductor process technology. A key issue is the level of granularity. Fine grain FPAA architectures (reconfigured at the transistor level, for example) will require more routing resources and will have more switches in the signal path than a coarser grain FPAA architecture (reconfigured at a macro-block level, e.g. integrators, S/H). However, the coarser architecture will be less versatile, i.e. it will be able to implement a narrower range of circuits than the fine architecture. Another issue is whether to make the CABs distinct for different circuit functions, or identical, but programmable to implement different functions. This choice will influence the area of the CABs and routability of circuits within the FPAA. CAB design can thus be seen to strongly influence the FPAA area, the routing requirements, the variety of circuits that can be prototyped and the performance of circuits prototyped on an FPAA. These issues were explored in a detailed study of CAB design, based on a set of application circuits for analog signal processing [18]. CABs of a few different types were designed; some of them were made reconfigurable where this could be done area-efficiently. The granularity level observed to be most area-efficient in this study was termed "building block," and included opamps, programmable resistors and capacitors.

1.2.4. Interconnect Architectures and Implementations. The choice of an interconnection architecture and its implementation will influence the routability of prototyped circuits and their performance. Analog circuits are far more sensitive than digital circuits to problems of fanout, noise, and the presence of switches in the signal path. Both hierarchical and full crossbar interconnection architectures have been used. Interesting work, influenced by Cellular Neural Networks, has been done in the area of cellular interconnect architectures, in which CABs are connected only to their nearest neighbors [3,10].

Pass transistors and CMOS transmission gates have been used as switches to reconfigure the topology of the circuit being implemented. Some approaches to implementing FPAAs have used circuits with MOS transistors operated in the sub-threshold region. In these implementations, voltage drops due to switch on-resistance are not a problem because of the extremely small currents in the circuits [16,19]. In linear circuit implementations of FPAAs, where the currents flowing through the switches are much higher, circuit techniques such as linearization of switch resistance [8] or judicious placement of transistor switches in the circuit embedding [9], have been applied to minimize the effects of the non-idealities of the switches. Other designs have made circuit structure reconfigurable without the use of extra switches in the signal path, by the use of discrete-time sampled-data techniques [3–5]. In these designs, switches used for circuit operation are also used to program the circuit structure. Alternatively, signals can be coupled from one CAB to another without reconfiguration switches in the signal path by changing the bias of an interface circuit such as a current source [10].

1.2.5. Programmable Components. Various methods have been used to implement programmable resistors. These have included the use of polysilicon resistors switched into circuits with pass transistors [20], complementary MOS transistor pairs with controlled gate voltages [21] and more complex transistor implementations of programmable resistive elements such as MOS transconductors [8]. Programmable capacitor arrays have been widely used, especially in switched-capacitor circuits [3,4], in which they can be made to emulate programmable resistors. The literature contains several publications and products in the areas of programmable amplifiers and filters [22–25]. On-chip tuning circuits may be required to produce bias voltages or currents that set the value of programmable components. These circuits often involve the use of digital registers to store component values, and digital to analog conversion to produce the bias voltage or current. .

1.2.6. Configuration Memory. Two types of configuration memory are typically found in a programmable analog integrated circuit. Digital registers are widely used to store the states of connection switches as well as the values of components such as programmable capacitor arrays. Analog memory is commonly used to store circuit parameters such as multiplication coefficients or the gate voltages of a MOS transistor in a circuit implementing a programmable resistor. Storing a voltage on a capacitor is the most common implementation of analog memory, and requires a means for refreshing the stored voltage [26,27], as well as consideration of errors due to charge injection from access transistors [28]. Error correction of the analog memory contents has been demonstrated [29], as has current-mode multivalued memory [30]. Organization of the digital memory has been done in different ways, including serially; in which all memory cells are connected together as a single shift register, serial-parallel; where a serial bit-stream is loaded into a horizontal shift register, the contents of which are then strobed into a row of configuration memory indexed by the bits in a vertical shift register, and finally random access, where address lines determine where the input configuration data goes. The latter two approaches can be used to reconfigure parts of the IC while other parts continue to operate.

It should be noted that configuration memory can occupy a significant proportion of the total die area of an FPAA integrated circuit.

1.2.7. CAD Tools. Field-programmable analog circuits to date have required configuration bit-streams ranging from several hundred to a few thousand bits to instantiate the circuits being prototyped on them. Generation of this configuration bit-stream from a definition of the circuit and downloading of the bits onto the chip needs to be handled by a CAD tool. Because of the many different circuit techniques used to implement existing programmable analog circuits, the use of many of the present CAD tools requires detailed knowledge of circuit design using the appropriate circuit techniques, or working at a higher level of abstraction, commonly called the macro-block level, with predesigned components from a library [31]. Some CAD tools display the entire FPAA architecture on a single screen, enabling a user to configure each block and make the interconnections that instantiate the circuit being prototyped [32–34].

The area of FPAA CAD tools has a number of research issues to be addressed. Software is required that will allow design and schematic capture in terms familiar to a circuit designer, followed by technology mapping, placement and routing for the target FPAA or FPMA architecture with adequate consideration of analog IC issues such as noise and layout parasitics. Recent moves towards the standardization of Analog Hardware Description Languages [35–37] should serve to unify efforts in CAD development.

1.3. Organization

This paper is organized as follows: Section 2 describes some early approaches to the design and implementation of non-monolithic configurable analog systems, and discusses some IC design approaches related to FPAAs and FPMAs. Section 3 looks at several first-generation works in programmable analog integrated circuit research. In Section 4 we describe the present state of the art, including academic and commercial research ventures, commercial products and patent literature. Conclusions and suggestions for future work are presented in Section 5.

2. Early Configurable Analog Systems and Related IC Approaches

By way of background, this section discusses some early configurable analog systems. Parametrically configurable building blocks and flexible interconnection systems were used to build analog signal processing systems that provided signal-conditioned and pre-processed signals to digital computers for further analysis. These approaches foreshadowed in intent later work that appears in monolithic form. We also discuss two CAD-compatible analog integrated circuit approaches that are related to FPAAs, namely metal-masked analog arrays and analog standard cells.

2.1. Analog Computers

In the 1960s analog computers were commonly used as hardware simulators in various areas of science and engineering. We describe a nuclear measurement system and a power system simulator based on systems of analog building blocks.

2.1.1. Arbel Nucleonic Computer. One of the first published attempts at identifying standard analog building blocks that could be used to build reconfigurable analog signal processing systems is presented in [38]. The application, in the field of nuclear instrumentation, was signal conditioning and preprocessing of the current-output signal from a radiation detector, before digital signal processing by a computer.

Current was chosen as the signal parameter for this system. The building blocks included linear amplifiers, discriminators, ADCs, time-to-amplitude converters, linear gates and analog memory modules. Where necessary, parametric configuration was performed by using front-panel controls. The blocks were interconnected with coaxial cables that could be used (by appropriate choice of the cable length) as signal delay elements.

2.1.2. Power System Simulators. Another early application of configurable analog circuits [39] sought to build a real-time simulator for power system studies, with a modular approach that would permit flexibility in configuring the parameters of the components and the topology of the system being modeled. To implement the simulator, circuit models of the power systems were created from differential equations in state variable form. The analog blocks used in the circuit implementation included integrators, linear amplifiers, non-linear elements, weighted summers and resistors. Power system components such as transformers, machines, transmission lines, filters and valves were modeled using this technique. Simulation studies were performed to model HVDC transmission and inrush transients of transformers.

2.2. FPAA Related IC Approaches

In this subsection we discuss two IC implementation approaches related to FPAAs, namely metal-mask programmable analog arrays and analog standard cells.

2.2.1. Metal-mask Programmable Analog Arrays. Metal-mask programmable gate arrays are widely used for short fab cycle-time, low-cost implementations of digital integrated circuits. The circuit being implemented is instantiated in a sea-of-gates (SOG) array using a custom set of metal-masks. In our discussion of field-programmable analog and mixed-signal arrays, metal-masked arrays are interesting as a related approach that seeks to define VLSI implementations of analog building blocks that can be used to build useful circuits. Metal-masked analog arrays are also important as an extension of the FPAA design flow as a smaller-area and lower-cost means of implementing, in high volume, circuits that have been verified using FPAAs/FPMAs. Such an extension is provided for IMP's family of EPAC integrated circuits [32].

Metal-masked array technology has long been in existence for bipolar designs [40]. To provide the functions required by analog designs, these arrays tended to have small numbers of coarse granularity cells that were more complex than the fine granularity devices present in large numbers in digital arrays. Special function cells such as high power devices and voltage references were also common. Interesting in the context of mixed-signal designs is a dense array manufactured by Ferranti Interdesign [40] that packed analog cells around a digital array and in the spaces between bond pads. Another innovation in the field of bipolar arrays was pioneered by Exar [41]; which developed device layouts that could be customized as either npn or pnp transistors by the use of a metal mask. A 1991 U.S. patent [42] awarded to Plessey Overseas Ltd. describes a cell for a semi-custom array that could be instantiated as an npn or pnp transistor, a resistor or a diode by use of an appropriate metal-mask. Recent work in bipolar analog arrays has resulted in a cell-based array for wireless applications

[43] in the GHz range. The array is composed of RF and digital cells that are parameterized and interconnected using a metal mask.

Duchene et al. [44] evaluated the use of existing CMOS digital SOG arrays for implementing analog circuits, and proposed an extended CMOS array consisting of the conventional sea of gates, along with features particularly suited to mask-programmed implementations of analog circuits. An experimental study [44] found that analog circuits implemented in an SOG array with transistors built out of serial and parallel combinations of unit transistors exhibited a 10–50% degradation in performance for parameters such as gain-bandwidth product, offset voltage and phase margin when compared to full-custom IC implementations. The performance attained was found to be adequate for many applications. To further increase the performance of SOG array implementations of analog circuits, floating wells, bipolar transistors and high-precision, high-value resistors are required. An extended array proposed in [44] added to the SOG array an ''analog field'' containing a bank of resistors, lateral pnp transistors and matched p-channel diffpairs in isolated wells. Area-efficient mask-programmable dedicated-function blocks, in the form of high voltage (100 V) output driver transistors, a bandgap reference and a low-power oscillator, were also included because of their high frequency of usage in analog circuits.

2.2.2. Analog Standard Cells. Analog standard cell design methodologies are more challenging to implement than their digital counterparts. This is because of the wide variations in specifications, such as bandwidth and DC or AC levels of signals, for different instances of a given analog building block (e.g. an opamp) that are used in different circuits, or even within a single circuit. Analog standard cell design methodologies therefore require cells that can be reused in different applications. This requirement is similar to that of FPAAs, although the number of different CABs used in an FPAA is typically much smaller than the number of cells in an analog standard cell library. A 1994 U.S. patent [45] describes a scheme for creating standard cells by connecting circuits that perform different functions through a standard interface that sets signal parameters such as DC bias and peak to peak levels.

Certain specialized classes of analog circuits have used standard cell design approaches. A library of low voltage analog and digital cells, and high-voltage and mixed-voltage analog cells was used to implement a 60-V, 10-A intelligent power switch in a 3 μm analog CMOS process [46]. High voltage cells included charge pumps, level-shifters and protection circuits, mixed-voltage cells provided the sensing functions in the interface between the high voltage circuits and the low voltage analog and digital control circuitry.

Analog circuits for RF and microwave communication have also been implemented using a standard cell technology. Operating in the GHz frequency range, building blocks such as voltage controlled oscillators, power amplifiers and RF switches have been implemented in a 0.8 μm BiCMOS process [47]. Especially interesting in the context of reconfigurable analog integrated circuits are the RF switches, which are typically used to share components such as filters, between the transmitter and receiver sections of a radio circuit [47]. The implementation of reusable building blocks, along with integrated circuit switches to connect them together will likely influence the development of field-programmable integrated circuits for these specialized applications.

3. Previous Work

Previous work in the area of programmable analog integrated circuits is dominated by programmable neural network ICs. Another more general approach is evident in circuit implementations of the multiplication of a signal vector by a matrix of coefficients. The first field-programmable analog arrays employed continuous-time subthreshold circuit techniques, using basic building blocks interconnected by transistor switches to implement prototyped circuits. A discrete-time approach to programmable analog circuits used controlled duty cycle chopping of signals followed by signal reconstruction to implement analog coefficient multiplication in z-domain filter circuits.

3.1. Programmable Neural Networks

Programmable neural networks are an important class of programmable analog circuits. A summary of important research and commercial works in the area of neural networks can be found in [48]. Hardware implementations of neural networks required programmable structure, adjustable gain in the neuron characteristic transfer function, and programmable interconnection weights. A block diagram of a reconfigurable neural network is shown in Fig. 3.

Programmable analog neural network research has yielded several commercial programmable neural network chips and systems [48–51], along with circuit techniques, IC implementations and experimental results for programmable analog functions and interconnect, that have been valuable to the development of more general programmable analog circuits. In particular, the storage of analog circuit coefficients on local capacitors refreshed with the results of a digital to analog conversion of the contents of digital memories [26,27], has provided means for parametric programming of analog circuits that have been used in other implementations of programmable analog circuits [52,53]. The core of the analog circuitry in one neural network processor [20] was a programmable resistor chip, containing five-bit programmable resistor networks used to implement the synapses. It is significant that the earliest monolithic implementations of programmable analog circuits, composed of basic undedicated building blocks [16,19], were designed with neural networks as important intended applications.

Cellular Neural Networks (CNN) were introduced in [54] as a new class of neural network circuits, characterized by local-only interconnections and spatially-invariant connection weights. Since the cells of a CNN are interconnected only with cells in a specified neighborhood, this class of neural networks is fixed in structure, but programmable by a set of templates which define interconnection strengths and bias constants for the cells. Research into locally-connected CNNs has led to the identification of other classes of analog circuits [10] that can be implemented on FPAAs with local-only interconnect architectures, and some FPAAs [3,10] have been designed with architectures that employ local interconnections to achieve programmable structure without the use of connection switches in the signal path. Programmable current-scaling circuits such as those used in [55,56] will allow the development of current-mode interconnect architectures. As analog circuits address problems in real-time image processing, where the density of "a computer at each pixel" might be required [1], programmable neural network research will likely continue to influence the development of configurable analog circuits. Research in the area of CNNs is progressing in an interesting direction [57], one that seeks to develop high level control of analog computing with the use of CNN based analog and logical computing units. VLSI implementations of these mixed-signal "analogic" processors will give further impetus to the development of programmable analog and mixed-signal integrated circuits.

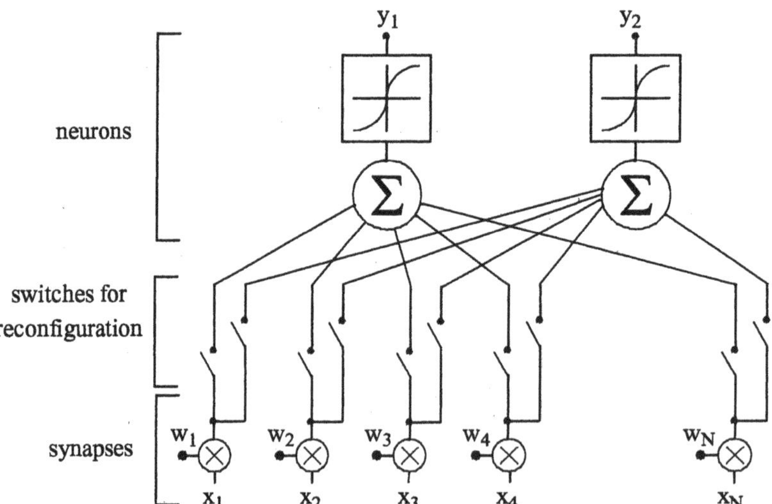

Fig. 3. A Reconfigurable Neural Network (from [28]).

3.2. *Programmable Analog Vector-Matrix Multipliers*

In [27], Kub et al. studied the vector-matrix multiplication operation

$$V_{Yi} = \sum_j W_{ij} V_{Xj}$$

where V_{Xj} is an input vector element, W_{ij} is a matrix of weights and V_{Yi} is the output of the multiplication operation. This operation finds application in many neural-network and signal processing algorithms. Fig. 4 shows a possible circuit implementation of this operation. An analog implementation of this operation offers the benefits of lower power consumption, higher density and faster performance than a digital implementation. A 32 × 32 programmable vector-matrix multiplier IC has been described in the literature [27], comprising an array of analog multipliers with weight coefficients in the analog memory periodically refreshed from the contents of digital registers. An architecture was described for implementing a multilayer neural-network from cascaded vector-matrix multipliers.

A 1991 U.S. patent [58] describes the design of a resistive network for the purpose of calculating, in real-time, a complex signal transformation of an input signal. The invention is targeted at complex signal transformations such as the Discrete Fourier Transform (DFT), which can be expressed as a sum of products, and hence as a matrix-vector multiplication operation. Time-domain samples of the voltage-mode input signal are created by passing it through a series of delay elements. These voltage samples are converted to currents by resistors in the network. The currents are then summed at network nodes, creating the output terms in the sum of products expression of the complex signal transformation being performed. The coefficients of the multiplication are thus inversely proportional to the resistance values.

3.3. *Sivilotti Proto-chip*

A field-reconfigurable IC called the Proto-chip, intended primarily for synthesis and test of analog neural-network architectures, is described by Sivilotti [19]. CMOS transmission gates were used as the active switch elements that connected basic resources such as differential pairs and current mirrors in a hierarchical routing network. A conceptual view of the Proto-chip depicting the circuit embedding of a transconductance amplifier is shown in Fig. 5.

While the on-resistance of connection switches was not an issue in implementing the low-current subthreshold circuits for which the Proto-chip was intended, parasitic capacitance of the wiring and switches in the interconnection network did present a problem. To minimize capacitance effects, transistor scaling and ''ring transistor'' layouts [19] were used to maintain a constant ratio of current drive to capacitive load for the transistors in the leaf cells. On board memory (SRAM) was used to store the state of each switch element, but no memory was provided for storing circuit coefficients.

Partial test results were published, including a ring-oscillator test of the performance of the interconnect

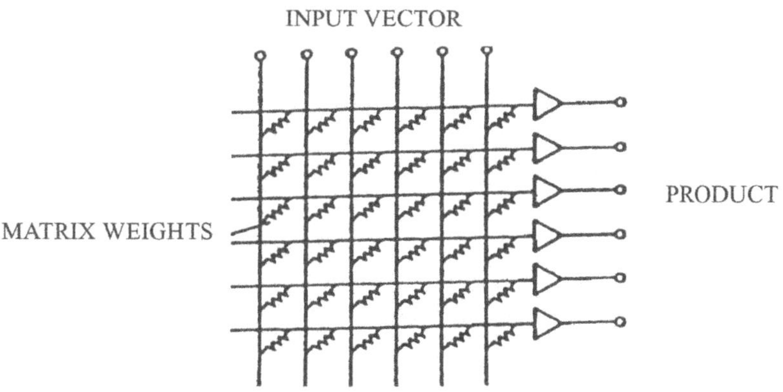

Fig. 4. Block diagram of a circuit implementation of matrix-vector multiplication (from [27]).

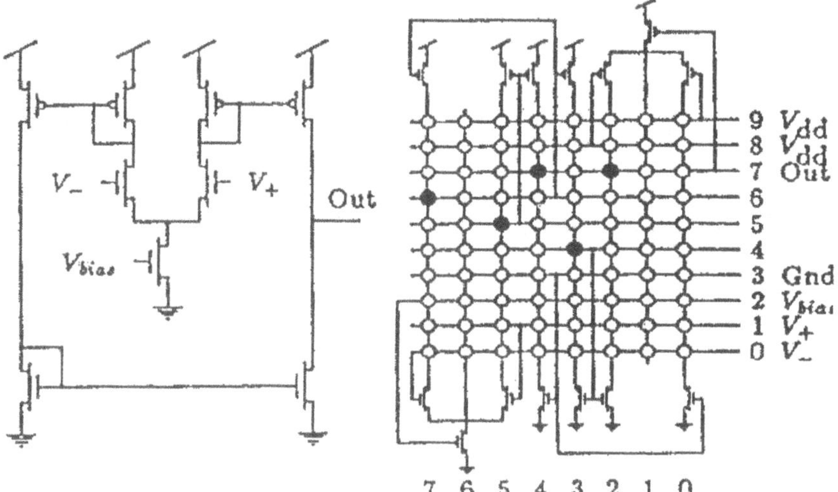

Fig. 5. Sivilotti Proto-chip showing circuit embedding of a transconductance amplifier (from [19]).

and measurements of the resistive and capacitive parasitics of the interconnect. A companion CAD tool to this FPAA system was a "metacompiler," which created a description of a Proto-chip given parameters for the interconnect and a leaf-cell layout which met some layout dimension constraints.

3.4. Lee-Gulak Sub-Threshold FPAA

In later work, Lee and Gulak [16,52,59] developed a low-power FPAA based on MOS sub-threshold circuit techniques. This IC employed both voltage- and current-mode circuits and was designed to implement structurally and parametrically reconfigurable neural-networks. Pass transistor switch networks controlled by SRAM-based memory elements were used as the active switch elements that connected basic resources such as differential pairs, current mirrors and transistors. Multi-valued memories were used to store circuit coefficients. The interconnect architecture of the FPAA was hierarchical. A circuit diagram of the CAB for this FPAA is shown in Fig. 6.

A prototype FPAA, comprising a sub-tree which included two Configurable Analog Blocks (CABs) and a switch-block, was fabricated. Experimental results verified the functionality of the FPAA and characterized the performance of the CABs in various circuit configurations. Die-to-die variations in subthreshold model parameters provided challenges to circuit operation in some applications. A further contribution of this work was the development of macro-models of the CABs to be used for the design and simulation of neural networks implemented on the FPAA.

3.5. Timing-Controlled ASP

Seeking to achieve the programmability associated with digital signal processing, without the need for complex circuit structures such as ADCs, DACs and microprocessors, an analog emulation of z-domain filters was proposed by Vallancourt and Tsividis [5,60]. For this class of filters, transfer functions are determined for a given filter topology by the signal gain in each network branch.

The implementation of these Analog Signal Processors (ASP) avoided the use of programmable resistor and capacitor arrays by chopping the input signal with a transmission gate switched at a digitally controlled duty cycle, followed by signal reconstruction to effect the analog multiplication that determines network branch gain. This technique, described in [5], uses a periodically reset integrator to reconstruct sampled signals and a Low Pass Filter (LPF) to reconstruct non-sampled signals, enabling all coefficients of the filter transfer function to be individually programmable using duty cycle control alone.

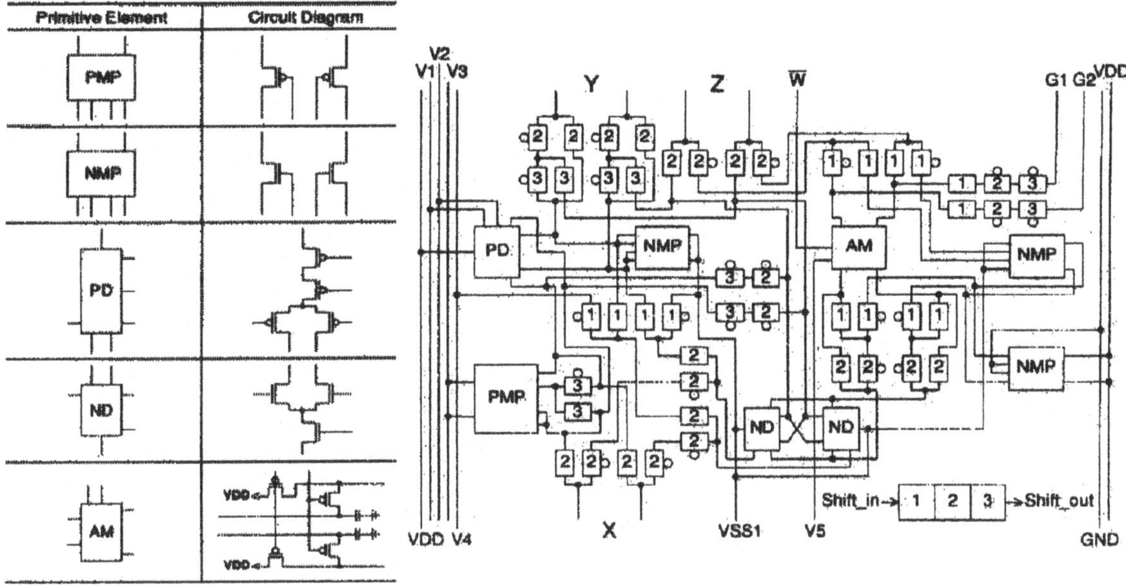

Fig. 6. CAB design for the Lee-Gulak sub-threshold FPAA (from [59]).

Weighted summation of signals could also be implemented by using this technique with appropriately sequenced chopping signals [60]. Programming of circuit structure was also demonstrated, with switch selection determining the filter topology, addressing the problem of interconnection network parasitics by using connection switches as circuit components. The timing-controlled technique has the added advantage of the capability to trade signal bandwidth for resolution in coefficient programming after fabrication, by varying the system clock and chopping signals. It should be noted that whether operating on sampled or non-sampled signals, the timing-controlled technique requires that input signals be bandlimited to at least one-half the chopping frequency.

Experimental results were published [5] for a sampled signal, fully-programmable biquad filter with a sampling rate of 16 kHz. A single second order section was used to implement an automatically reprogrammed, time-interleaved, fourth order low-pass transfer function. In addition to being a possible candidate for FPAA architectures, this technique represents a CAD-compatible analog design method appropriate for circuit implementation on FPMAs, in which the digital circuitry would generate the timing signals, and the analog circuitry would implement the switches and LPFs used in the ASP.

4. Present State of the Art

In this section we describe various field-programmable analog and mixed signal integrated circuits that have resulted from research at academic and commercial laboratories, as well as commercially available programmable analog components and integrated circuits. This is followed by a discussion on the status of intellectual property ownership in this field. A status table and a time-line of major achievements in the fields of FPAA and FPMA integrated circuits are presented at the end of the section.

4.1. Field-Programmable Analog Arrays

The body of present work consists of several field-programmable analog array integrated circuits that have resulted from research in the academic and commercial sectors.

4.1.1. PMeL-Motorola FPAA. Pilkington Micro-Electronics (PMeL), has published details [3,61,62] of an FPAA based on switched-capacitor circuit techniques. This IC consists of a 4×5 array of programmable analog cells each containing an operational amplifier, a programmable capacitor array and a set of CMOS transmission gate switches. Local RAM within each cell holds digital configuration data that is decoded by a configuration manager to set the connectivity and switch phasing of the switches in the cells, implementing the desired function. Fig. 7 shows a conceptual view of a cell implementing a switched-capacitor integrator. Each cell is locally interconnected with nine neighboring cells, with each cell selecting at its input the signals required to implement the programmed function. Global interconnections are facilitated by a pair of routing tracks in each horizontal and vertical channel in between cells, with an array of switches at the intersections of tracks to make cross connections. Parasitic-insensitive circuit techniques were used in the design of the switched-capacitor cells to minimize the effects of signal degradation in the FPAA routing. The cells can be configured to implement several analog subcircuits including operational amplifiers, comparators, gain stages, first-order filter sections, integrators and differentiators [31]. Switches can be dynamically controlled by signals internal to the array, allowing the implementation of designs such as switched-capacitor full-wave rectifiers. A circuit embedding for a PCM CODEC using several analog cells was described in [3]. To date, experimental results have been published for a test of the analog cell as a variable gain amplifier. Commercial products based on this research are expected to be released in 1997 following Motorola's recent acquisition of PMeL [63].

4.1.2. Lee-Gulak Transconductor based FPAA. A MOS-Transconductor based FPAA has recently been described in the literature [8,53]. It consists of operational amplifiers and programmable capacitors linked by a transconductor based interconnection array. The innovation in this design is that the switches in the interconnection network are in fact programmable linear resistors in the circuit being prototyped. Fig. 8 shows the circuit embedding of a filter biquad section in this FPAA. Programmable

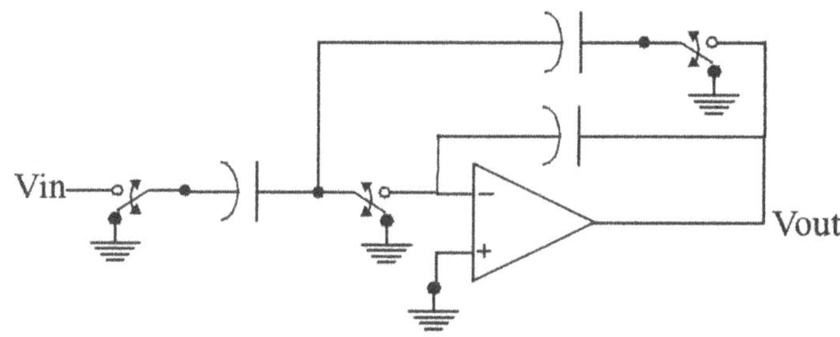

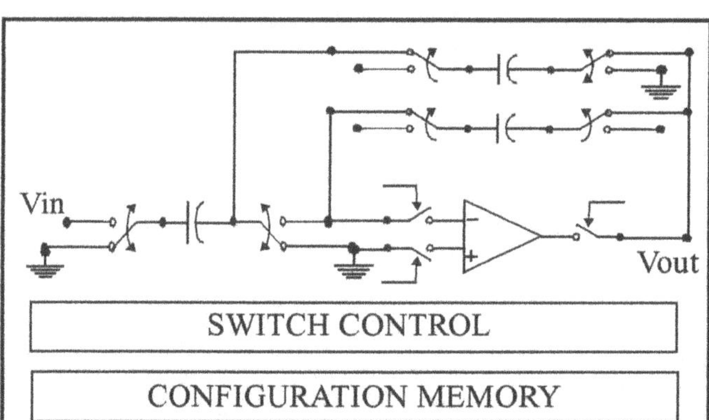

Fig. 7. Conceptual view of the PMeL-Motorola switched-capacitor FPAA cell showing embedding of an integrator.

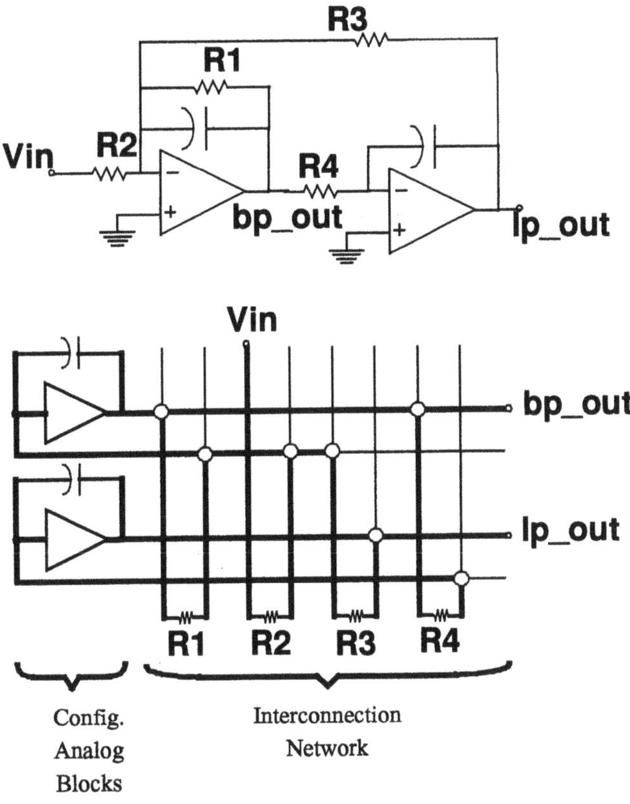

Fig. 8. Lee-Gulak FPAA showing circuit embedding of a filter biquad.

resistors are implemented as four-transistor MOS transconductors [64], a circuit arrangement that uses four matched MOS transistors operated in the linear region, cross-coupled such that the non-linear components in the drain currents are cancelled for fully-differential voltage inputs and current outputs. This results in a circuit that can be used as a programmable linear resistor, a signal controlled resistor, a signal multiplier or a polarity change switch. A modification [18] of the basic transconductor circuit, shown in Fig. 9, splits each transistor into three serially connected transistors, with the transistors in the centre implementing a high resistance value set by the voltage contents of an analog memory periodically refreshed from on-chip 10-bit digital registers, and the transistors on the ends implementing lower-resistance termination parts that switch the resistor into the circuit being implemented by the FPAA.

This FPAA was designed to implement applications in the area of analog signal processing. An analysis [18] was carried out for FPAA implementations of a set of benchmark circuits in this field, with functional blocks reconfigured at five different granularity levels, from the transistor level (NMOS, PMOS) to the sub-system level (S/H, ADC, DAC). Parameters studied included the versatility of the blocks at each level (a measure inversely proportional to the number of distinct blocks required to implement a benchmark circuit), the ratio of usage between components (how many resistors were present in the benchmark circuits for each instance of an opamp), and the routing resources required for each granularity level. The building-block level (e.g. opamp, MOS transconductor, capacitor, diode) emerged as a good candidate for FPAA granularity because of the high versatility of the blocks, the area efficiency and moderate routing resource requirements when compared to other granularity levels for implementations of the benchmark circuits.

A fully functional prototype, fabricated in a 1.2 μm CMOS process, was reported in [53]. Functionality and programmability were verified for the configurable analog blocks and interconnect, and several

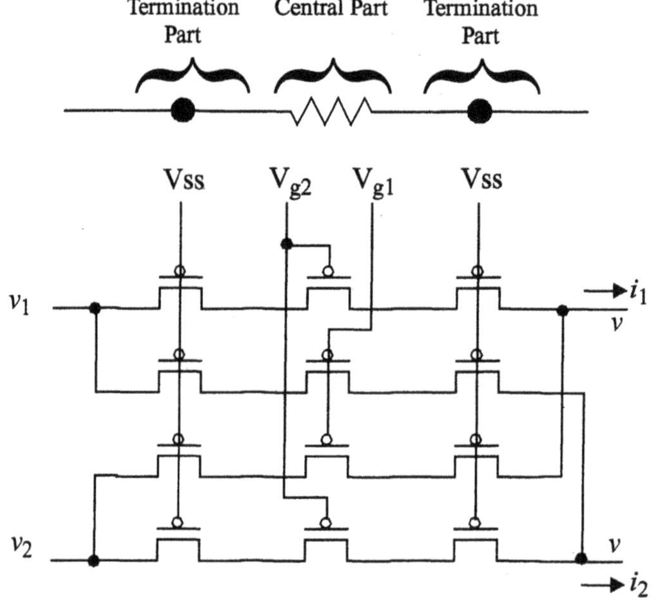

Fig. 9. Modified MOS Transconductor (from [53]).

application circuits operating in the audio-frequency range were implemented and tested. A PC-based CAD tool [33,34] was used for schematic capture, configuration bit-stream generation and device programming.

4.1.3. Analogix bipolar current-mode FPAA. A current-mode bipolar FPAA intended to operate at hundreds of MHz is presented by Analogix Corp./ Portland State University in [10,65]. Inspired by the success of locally-connected architectures of Cellular Neural Networks, an analysis of analog circuits showed that certain classes of circuits, notably multistage amplifiers, biquad filters and ladder filters, are readily implemented on an FPAA with a local interconnect architecture. The resulting FPAA comprises an array of homogenous analog cells, each locally interconnected with four neighboring cells without the use of switches in the signal path.

The cells use fully-differential current-mode circuit techniques to implement a set of mathematical functions. The general form of the functions is

$$Y = k \cdot \left[\sum_{w_i \in W_1} w_i X_i \right] \cdot \left[\sum_{w_j \in W_2} w_j x_j \right]$$

where X is the set of inputs to the cell, W_1 and W_2 are independent sets of programmable weights and k is a programmable gain factor. Special cases of this form implement the operations of summation and multiplication. The result Y can then be integrated with a programmable loss-factor, and passed through a programmable non-linear thresholding and clipping circuit. A block diagram illustrating the programmable cell functions is shown in Fig. 10. Each cell contains a control block which stores cell configuration data, and provides programming signals to the circuits in the cell to control its operation. The control block is capable of configuring the function of the cell depending on signals internal to the FPAA, allowing it to implement such functions as minimum/maximum signal tracking. An additional level of global interconnect was provided to increase the flexibility of the FPAA, however, its use is intended to be minimized because it adds switches to the signal path. Circuit embeddings have been demonstrated for an eighth-order elliptic bandpass filter, a circuit which tracks the solution of a system of linear equations, multi-valued logic applications and a fuzzy-logic controller.

4.1.4. Multi-Function Signal Detection Block. Chang et al. [9] describe an approach to the construction of an FPAA based on a collection of "Multi-function Blocks," each of which implements

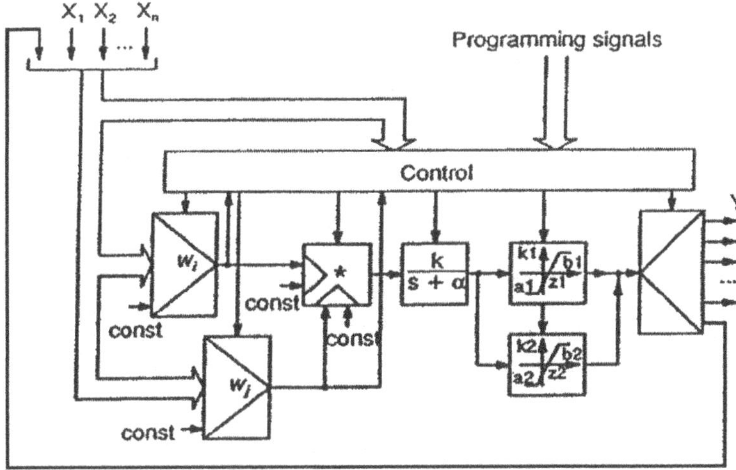

Fig. 10. Block diagram of a cell in the Analogix FPAA (from [65]).

a set of the various analog functions required by the class of analog circuits being prototyped. The function of each block is determined by the state of a number of switches within the block which make the circuit reconfigurable. A "Multi-function Signal Detect Block," or MFSDB, containing an opamp, a diode, a capacitor and a set of CMOS transmission gates, can be configured as a voltage follower, a peak detector, a half-wave rectifier and a comparator with variable (or no) hysteresis. The block is designed so that, as much as possible in the implementation of the above circuits, the resistance of the non-ideal switches appears either in low-current paths such as those leading to the high-impedance inputs of the opamp, or in series with the opamp output, inside the feedback loop. In the former case, the voltage drop across the switch will be kept small, in the latter the effect of the increase in the opamp output resistance due to the switch can be eliminated by following the MFSDB with a buffer stage. A characterization cell was fabricated in a 2.4 μm CMOS process. Switches and interconnect were reported to occupy a significant percentage (65%) of the layout. Experimental results verified operation of the cell in its various modes, for signals up to 1.2 MHz for the voltage follower, and 11.2 kHz (opamp slew-rate limited) for the half wave rectifier. The sampling time of the sample-and hold circuit was 1.6 μs, response time for the comparator was on the order of 3 μs. The effects of interconnect on the performance of an FPAA based on this cell were not addressed. Future work was to focus on building more "Multi-function Blocks," for functions such as filtering and input/output.

4.1.5. Kutuk & Kang Switched-Capacitor FPAA. A recent publication [4] describes the design of an FPAA based on switched-capacitor techniques, designed to operate at frequencies up to 125 kHz. Each CAB in the design contained a lossless integrator and a lossy integrator, implemented using stray-insensitive switched-capacitor techniques and connected in a loop using switched capacitor connections as shown in Fig. 11. The CAB could be configured using three schemes. By interchanging switch phases, inverting or non inverting integration could be achieved. Switches could be used to connect or disconnect CAB elements from the circuit. Finally, capacitor values could be programmed to one of four capacitance values. The interconnection network between CABs in the FPAA was also composed of switched capacitors configurable in the same ways as in the circuits within the CAB, allowing the use of switched capacitor interconnections to implement signal summation at a node and the use of unswitched capacitor interconnections to realize finite transmission zeros. An Algorithmic State Machine execution unit was designed to configure the array from configuration bits stored in an EPROM. Eight-bit control words were decoded by the control circuit to produce the appropriate control signals for the CAB and interconnection network. The frequency of the clock for the switched-capacitor circuits was pro-

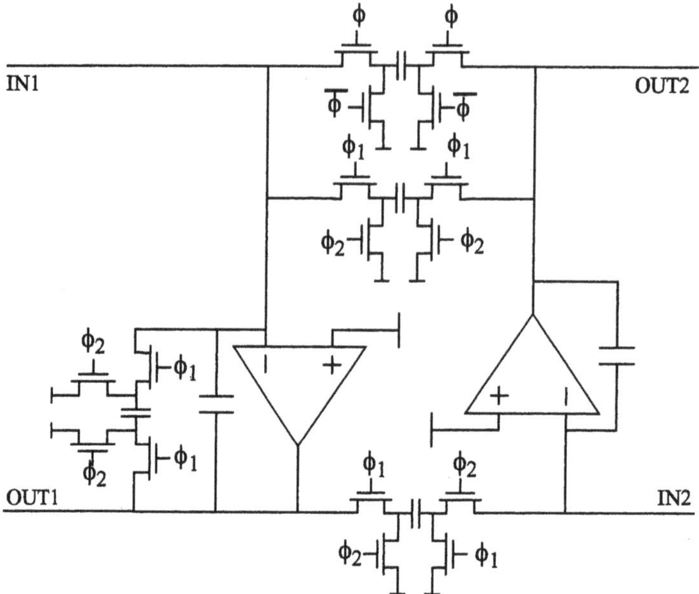

Fig. 11. Switched-Capacitor CAB (from [4]).

grammable in multiples of two between 2 kHz and 256 kHz. Simulation results were given for circuit embeddings of a third-order low-pass elliptic filter, a fourth-order bandpass filter, a balanced modulator and a quadrature sinusoidal generator.

4.1.6. Recent Current Mode Approaches. FPAAs based on current mode approaches have been described in the literature recently. Premont et al. [11] designed an FPAA with CABs consisting of two current conveyors, along with tunable resistors and capacitors. A discrete-time current mode approach was used in the design of a switched current FPAA by Chang et al. [7]. Embabi et al. [12] describe an FPAA based on a folded cascode integrator circuit, which can be configured to function as an integrator, an amplifier or an attenuator. Undoubtedly we will see many more innovative ideas in this area.

4.2. *Field-Programmable Mixed-Signal Arrays*

To date, field-programmable mixed analog-digital integrated circuit designs have been based on the union of previously designed analog and digital arrays, along with some provision for the exchange of signals between the two domains. Two options, illustrated in Fig. 12, exist in this respect; dedicated data converters (Fig. 12a) can be implemented to perform the analog to digital and digital to analog conversion of signals, or the converters can be built out of analog and digital resources available in the arrays (Fig. 12b). The works discussed in this section include examples of both options.

4.2.1. U. Toronto MADAR FPMA. A continuous-time FPMA prototype IC called MADAR was described in [66]. It allowed on-chip exchange of signals between the analog and digital domains. This array was built from a previously designed digital FPGA based on four-input lookup tables [67] and the continuous-time FPAA from [53]. Dedicated data converters were designed for the interface between the FPAA and FPGA with a view to improving area efficiency and performance over an interface constructed out of resources within the analog and digital arrays. Several mixed-signal circuits with applications in data conversion and signal processing were studied, to determine the sizes of the analog and digital arrays and the characteristics of the interface in between them necessary to design a prototype chip. The results indicated that the number of interconnections between the analog and digital domains was small compared to the number of connections within each domain, so the

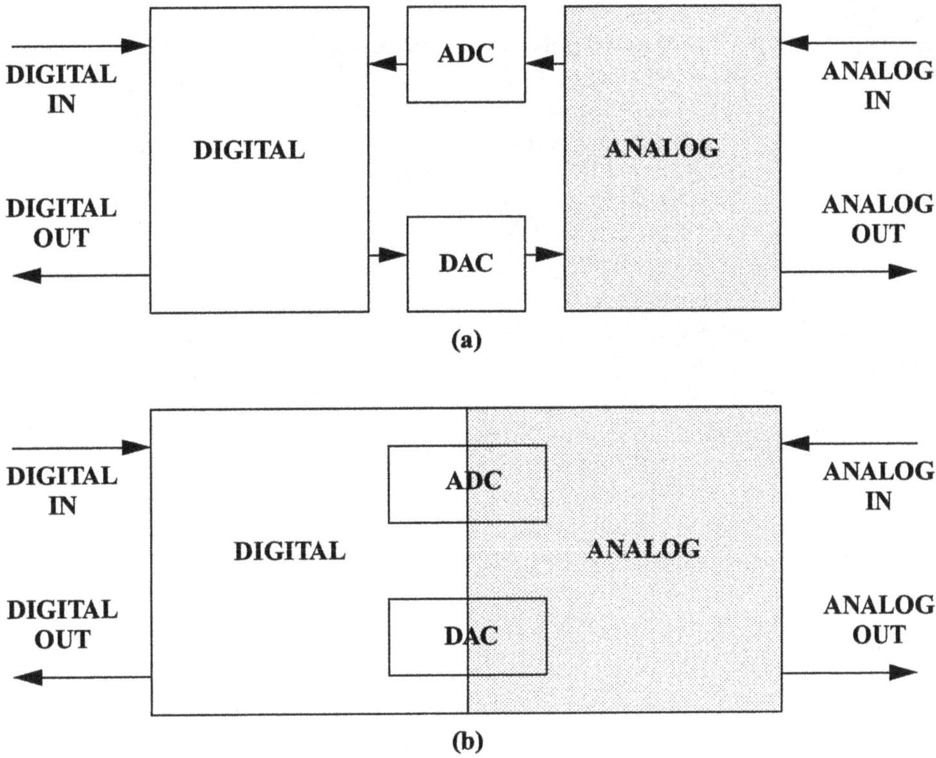

Fig. 12. FPMA Conceptual block diagrams (from [66]).

data converters in the interface were directly connected to the routing networks of the FPGA and FPAA.

Configurability was designed into the converters as a digitally programmable trade-off between converter resolution and the number of converters. The digital to analog converters were based on a MOS current division technique [68], and made programmable with a single configuration bit to split a 4-bit DAC into two independent 2-bit DACs. The analog to digital converter design exploited the fact that multi-step converters are built out of multiple lower-resolution converters. A single configuration bit was used to enable two 2-bit converters to be used independently, or to form a single 4-bit, two-step ADC. The interface included direct connections between the analog to the digital domain, extending the interface to permit analog comparator outputs to tie into the digital routing network, and to allow digital signals to control analog switches in the analog array. A circuit embedding and simulation results were given for a dual-slope analog to digital converter.

4.2.2. PMeL-Motorola FPMA. PMeL has published details of an FPMA [31,69,70] resulting from the integration of their switched-capacitor FPAA [3] with a previously-existing digital FPGA layout, to which the FPAA layout was pitch-matched. Limited details were given concerning the interface between the analog and digital domains, but the design criteria given for the interface describe the exchange of data between the domains via data converters constructed from the analog and digital programmable resources, and dynamic control by FPGA circuits of cell-configuration, component values and reference voltages in the FPAA section. Mixed-signal applications were targeted in the areas of digitally tunable filters, phase-locked loops, A/D and D/A converters. Experimental results were not published. Design of circuits for the PMeL arrays is performed on a CAD tool which supports design at the macro-cell and switched-capacitor circuit levels.

4.2.3. Faura et al. FIPSOC. The design of a FIeld-Programmable System On Chip (FIPSOC) is described by Faura et al. in [71]. The proposed IC is to contain a coarse-grained lookup-table based FPGA, reconfigurable analog blocks and an 8051 microprocessor core. The device is targeted at reducing the design time for systems comprising digital, analog and software components. The analog blocks on the chip each contain four channels with programmable settings for gain, filtering and comparison. A data-conversion block contains Analog to Digital and Digital to Analog converters with reconfigurable resolution. A routing network was provided for the analog systems. The interface between analog and digital systems allows outputs from the ADC blocks and comparators to tie into the digital routing networks, and the outputs of the FPGA can drive the inputs of the DAC blocks. In addition to these interactions, signals within the analog block can be read by the microprocessor through the ADC. This design features a second set of circuit configuration memory which can be used to instantaneously reconfigure all or part of the device by issuing a command to the microprocessor.

4.3. *Commercial ICs*

Several commercial ICs are of interest in our discussion of FPAA building blocks. They include a simple set of analog building blocks and flexible interconnect on a single chip, simple programmable analog components such as digital potentiometers, programmable amplifiers and filters, and more highly integrated systems for multi-channel analog control, signal conditioning, and data acquisition applications. We begin by discussing the GAP-01, followed by programmable analog components, leading up to the first modern programmable analog integrated circuits available commercially.

4.3.1. GAP-01. The GAP-01 [72] shown in Fig. 13, released in the early 1980s by Precision Monolithics Inc., was an early attempt by industry to monolithically implement a set of analog building blocks that could be structurally and parametrically configured to implement different applications. The building blocks available on this IC included two transconductance amplifiers, the outputs of which could be selectively connected, in a patented arrangement [73], to a single output buffer by means of a pair of on-chip, digitally-controlled low-glitch current-mode switches. The integration of digitally-controllable interconnect and the use of special circuit techniques to compensate for switch non-idealities mark important achievements in the evolution of field-programmable analog ICs. An uncommitted com-

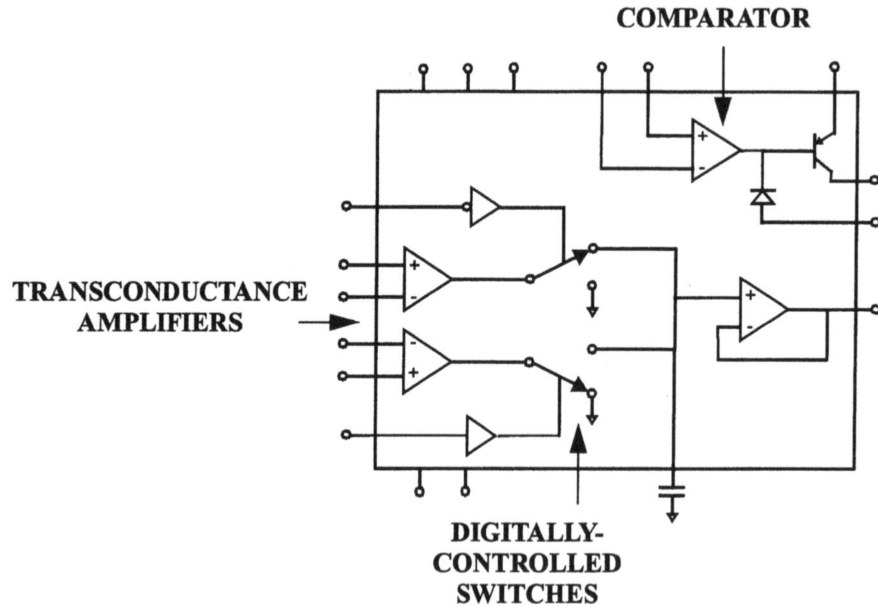

Fig. 13. The GAP-01 General Analog Processor (from [72]).

parator was also available, with its high-level output voltage parametrically set by the use of external resistors. The GAP-01 datasheet cites applications such as a two-channel sample-and-hold amplifier, an absolute value amplifier and a successive-approximation analog-to-digital converter.

4.3.2. Digital Potentiometers. Also relevant to our discussion of programmable analog circuits are digital potentiometers (Fig. 14). Xicor [74] introduced this type of ICs in 1990. Intended to replace manual potentiometers in circuits that require trimming, these ICs come in 1 kΩ and 10 kΩ sizes, with the wiper position controlled by a digital UP/DOWN counter which, in turn, is controlled by a human-push-button interface. Wiper positions can be changed in linear or logarithmic steps. Non-volatile EEPROM storage is available to retain the wiper position.
Analog Devices family of digital potentiometers [75] are designed for control by digital circuitry. A 3-pin serial interface loads an 8-bit number to set the wiper position of 10 kΩ, 50 kΩ or 100 kΩ potentiometers. These devices have a bandwidth of 600 kHz, and can be used to implement programmable filters and time constants as well as in circuit trimming applications.

Efforts in the development of such integrated circuits will bring commercial products that add digital programmability to analog circuits at the board level, and will result in improved integrated circuit implementations of analog circuit components, programmable over a large dynamic range, that will be essential for the design of future FPAA and FPMA ICs.

4.3.3. Programmable Gain Amplifiers. Burr Brown makes a digitally-programmable instrumentation amplifier [23], using two configuration bits to vary the gain in decade (1, 10, 100, 1000) or binary (1, 8, 64, 512) steps for applications in data acquisition and instrumentation. The gain is set by a resistor network with resistors selected by switches placed in series with the high-impedance amplifier inputs, minimizing gain errors due to the switch ON resistance. To minimize gain errors due to resistance variations with temperature, the resistor network is fabricated on a separate substrate from the amplifiers. The 3 dB bandwidth of the instrumentation amplifier varies inversely with the gain setting, between 2.4 kHz and 500 kHz.

Analog Devices manufactures a high-bandwidth

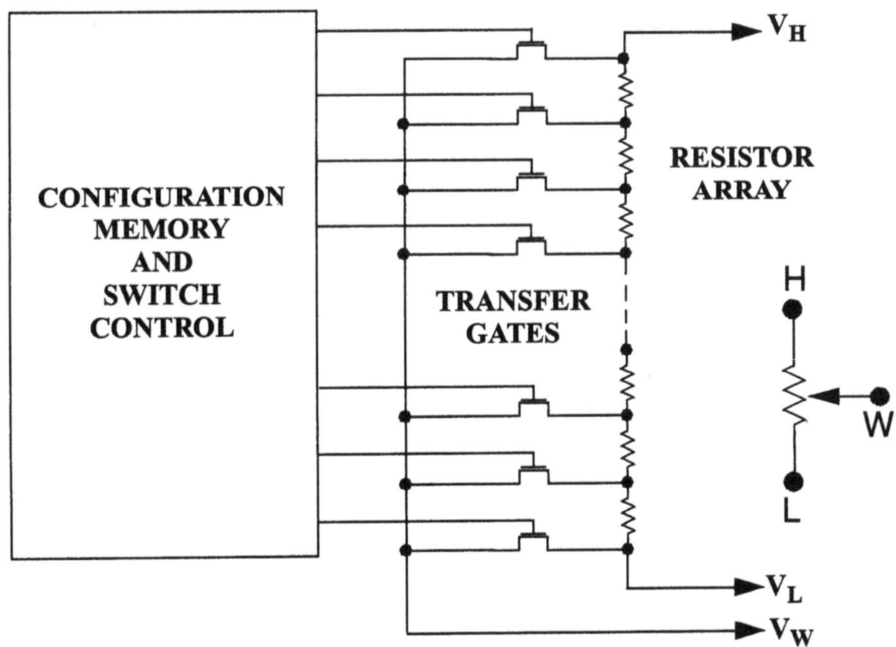

Fig. 14. Block diagram of a digital potentiometer (from [74]).

variable gain amplifier [22], for RF/IF and video automatic gain control and measurement applications. This device consists of a variable-attenuator input stage implemented as a seven-stage R-2R resistor ladder, followed by a fixed gain negative feedback amplifier. The attenuation between tap-points on the R-2R ladder is 6 dB; a proprietary circuit technique is used to interpolate continuously between tap-points based on a control voltage input.

4.3.4. Programmable Filters. To demonstrate some of the work in this area, we present two commercial approaches to programmable filter design from the numerous relevant publications, products and patents in the area.

The Universal Active Filter [24], manufactured by Burr-Brown consists of a biquad section built out of opamps and on-chip resistors and capacitors. A few off-chip resistors are required to use this device to implement Butterworth, Chebyshev and Bessel all-pole filters. An uncommitted opamp on the chip allows implementation of other filter topologies such as the Inverse-Chebyshev. A CAD tool solves filter design equations to calculate the values of the external components and display filter performance specifications. Continuous-time filters with up to 100 kHz band width can be implemented using this chip.

A versatile continuous-time programmable filter manufactured by IMP for applications in the read channel circuits of magnetic tapes and disk drives is described in [25]. The filter architecture was composed of four independently programmable cascaded sections. Two were first order sections and the other two were second order sections. The RC time constants of the opamp integrators in the filter sections were implemented using oxide capacitors and triode-region MOSFETs. An automatic tuning circuit used a phase locked loop and an external frequency reference signal to set the frequency of an internal oscillator. The voltage thus developed on the gate of a triode-region MOSFET in the oscillator was then used to bias the MOSFET in the filter so that the filter cutoff frequency was set to a digitally-programmable fraction of the input reference frequency. A set of 8-bit R-2R ladder DACs allowed further control of the filter transfer function. A three-wire serial interface was provided to load the 18 eight-bit digital configuration registers that program the filter. The cutoff frequency of each section was programmable from 100 kHz to 4 MHz.

4.3.5. Texas Instruments Analog Interface Chip. TI offers a reconfigurable Analog Interface Chip (AIC) in the TLC3203X/4X [76]. The IC functions as the analog front/back end interface to a DSP in applications such as modems, speech processing, industrial process control and biomedical applications. A continuous-time anti-aliasing filter precedes the switched capacitor analog circuitry. The gain of the analog pre-amplifiers can be varied by changing the bit pattern in a control register, effecting software controlled signal conditioning. Digitally controlled analog multiplexers handle multi-channel operation of the device. Reconfigurable analog circuitry is also used to add testability to the chip; a loopback feature connects the analog output to the analog input. Under the control of a DSP, digital words can then be loaded into a register, converted to an analog signal by the on-chip DAC, then reconverted to digital form by the ADC for comparison to the original words to verify operation of the converters.

4.3.6. IMP Switched-Capacitor FPAA. In 1994, IMP Inc. introduced the 50E10 Programmable Analog Signal Conditioning Circuit (Fig. 15), the first IC in their series of Electrically Programmable Analog Circuit (EPAC) devices [13,32,77]. Aimed at multi-channel analog signal conditioning applications that require channel-dependent signal scaling and offsets, the 50E10 has an analog multiplexer front end, variable offset and gain modules, a summing amplifier and output blocks that can be configured as comparators, amplifiers or sample and hold circuits. Virtually all the analog blocks on the IC employ switched-capacitor circuits and on-chip anti-aliasing input filters and output filters are included in input and output blocks. An uncommitted opamp is provided for wiring into the circuits using external connections. On-chip EEPROM is available to store user-programmable circuit configurations. In-system reprogramming of the chip is also supported.

Possible applications of this device are sensor signal conditioning, process control, data logging, DSP front ends and test equipment. At unity gain, the chip's opamps support a stable, large-signal bandwidth of 15 kHz with the internal antialiasing filters turned on. The design goal for the EPAC was to sacrifice circuit configuration flexibility in favor of performance and ease-of-use of the device within the targeted application area. In particular, the design seeks to insulate users from analog integrated circuit

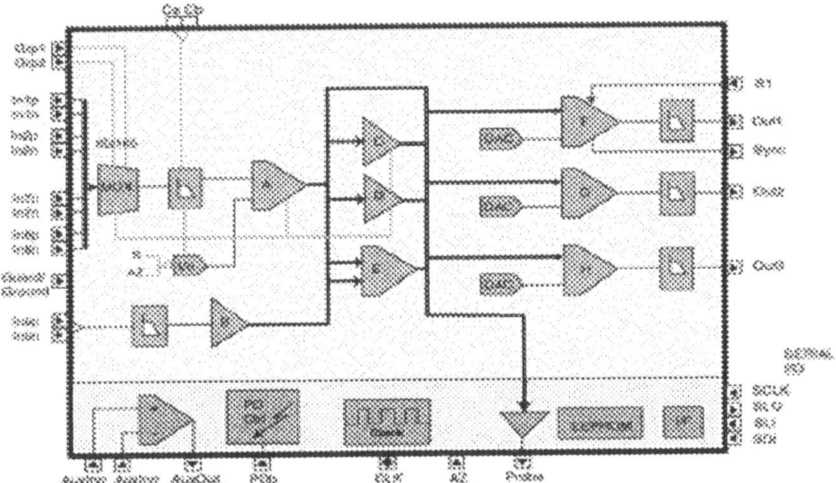

Fig. 15. Block diagram of the IMP EPAC 50E10 (from [32]).

design issues of temperature matching, offset-cancellation, parasitic coupling, loading effects, gain-bandwidth trade-offs, stability and power management. Limiting the range of circuit topologies that could be implemented in the device allowed optimization of building block and interconnect characteristics to achieve the above goals. A PC-based CAD tool is used for schematic entry, configuration bit generation and programming of the EPAC.

Follow-on products to the 50E10, with added functionality, announced recently by IMP [32] continue to target signal conditioning applications. The 50E20 Programmable Gain and Function Amplifier will add serial-output, 8-bit analog to digital conversion capability to a signal path similar to that of the 50E10, using a successive-approximation ADC. The 50E30 Programmable Monitoring and Diagnostic Data Acquisition IC will build on its predecessor with the implementation of programmable monitoring of multiple channels for threshold or window specifications, and can be set up to interrupt a microcontroller when a channel registers an error, provide a serial-output status word indicating the error channel and type, and be reprogrammed by the microcontroller to output an 8-bit measurement of the channel on which the error was detected.

4.3.7. Adaptive Logic Fuzzy Logic Controller PAIC. Adaptive Logic [6] produces the AL220 advertised as an analog microcontroller featuring a programmable analog IC (PAIC) capability. The device operates on a set of four analog input signals, which are processed according to a set of fuzzy control rules to produce four analog outputs. The analog inputs are converted to 8-bit digital values by on-chip analog to digital converters, then processed by a digital core programmed by a 256×8 EEPROM. The 8-bit digital outputs from the core are converted to analog outputs by a digital to analog converter followed by a sample and hold circuit. Fig. 16 shows a block diagram of signal flow in the PAIC. The maximum processing speed of the digital core is limited by data sampling and processing operations to 10 kHz, at the maximum clock frequency of 10 MHz. A CAD software tool is used to convert the fuzzy control rules to a bit stream which configures the device. Applications in fuzzy logic control are demonstrated for this IC in [6].

4.3.8. Zetex TRAC. The first commercial continuous-time FPAA was announced by Zetex in 1996 [62,78]. The Totally Reconfigurable Analog Circuit (TRAC) consists of 20 cascaded analog cells, each of which can be configured to implement one of six functions, namely Add, Negate, Pass, Log, Antilog, Rectify. In addition, a cell can be turned off or programmed to be a standalone op-amp for use with external components. On-chip interconnections are local only, with the output of each cell connected to the input of the next. The inputs and outputs of all cells are brought off-chip to enable global routing

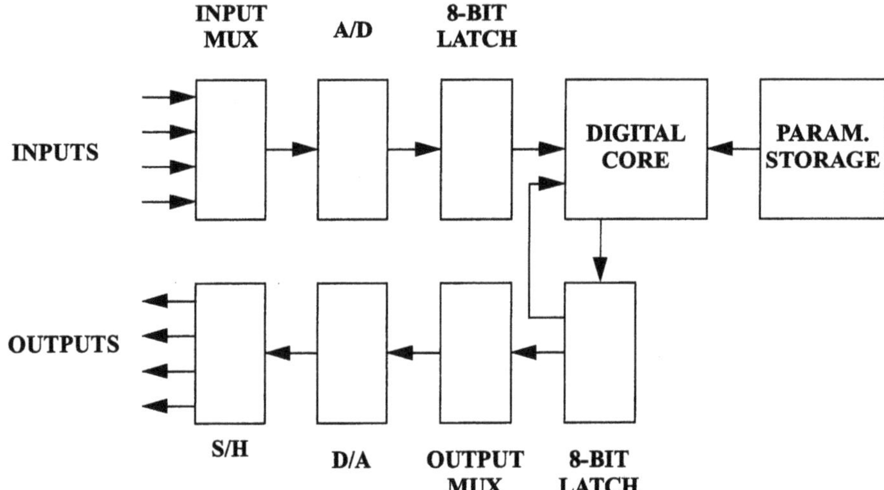

Fig. 16. Block diagram of signal flow in the Adaptive Logic PAIC (from [6]).

using external connections and/or components. A shift register provides volatile storage of circuit configuration data.

4.4. Intellectual Property Status

A review of U.S. patent literature in the area of programmable analog and mixed signal circuits reveals several patents, indicating significant commercial potential in this area. As early as 1977, a patent was granted for a digitally controlled variable conductance [79], composed of a fixed resistor in parallel with a series combination of another fixed resistor and a MOS transistor with its gate voltage controlled by a digitally controlled variable duty cycle generator. The low-glitch current-mode switch employed in the Precision Monolithics GAP-01 [72], was patented in 1981 [73]. In 1987, Czarnul was granted a patent [80], for the MOS Transconductor, a field-programmable conductance element used in [53] and in other applications such as programmable filters. In an arrangement which might be applied to FPAAs, analog standard cell methodologies or analog metal-masked arrays, Kabushiki Kaisha Toshiba's 1994 patent [45] describes a scheme for constructing standard cells from existing analog integrated sub-circuits and standard input–output interfaces which allow different analog subcircuits to be interconnected with standard DC bias and signal peak-to-peak levels.

Pilkington Micro-Electronics' 1993 patent [21] describes the design of an FPAA consisting of an array of operational amplifiers, programmable resistors constructed from multiple pairs of complementary MOS transistors and programmable capacitors with their range extended, at the expense of Q-factor, by an impedance-multiplication technique. The architecture described in this patent, along with switched-capacitor circuit techniques described in a 1994 UK patent [61], forms the basis of the PMeL-Motorola FPAA and FPMA designs. Kawasaki Steel describes, in a 1994 patent [81], an FPAA design consisting of operational amplifiers, passive resistor and capacitor elements interconnected with pass transistors. The intellectual property in this field continues to grow. Analogix [10], has a patent application pending for their bipolar FPAA design, as does Xicor, for their non-volatile digital potentiometer [74].

One of the first steps towards the evolution of an FPMA can be found in a 1992 Actel patent [82], that describes a user-programmable integrated circuit with configurable analog and digital circuit modules and interconnection networks and programmable ADC and DAC interface blocks. Patents granted to AMD in 1992 and 1993 [83,84], describe two types of PLAs in which an analog front-end or a back-end is married with a programmable logic array. A programmable mixed analog-digital, multi-channel signal-conditioning integrated circuit is patented by Hewlett-Packard [85], for applications in biomedical measurement.

4.5. FPAA/FPMA Status

The status of major research and commercial works in the area of FPAAs and FPMAs is summarized in Table 1. A timeline depicting the sequence of important achievements in the area of programmable analog circuits is shown in Fig. 17.

5. Conclusions and Future Work

The evolution of user-discretionary analog systems has shown notable progress in the last five years. While it is still relatively young compared to the digital FPGA industry, a survey of the area of programmable analog integrated circuits from historical and current perspectives has shown it to be active and diverse. Several indicators point to continued progress and development, namely the emergence of special sessions on FPAAs in conferences on programmable devices, the existing portfolios and continued assembly of intellectual property in the area by several corporations, the packaging of analog design software with an associated FPAA IC in the recent IMP and Zetex products, an FPGA industry with an expanding financial base searching for product differentiation, the trend in the IC industry towards complete systems on a chip, and increasing

Table 1. Status of Field-Programmable Analog and Mixed-Signal Integrated Circuits.

Group	Circuit Technology	CAB Verified	Intercon. Verified	Applic. Verified	CAD Tool	Patent Issued	Product Available	FPMA Extension
Vallancourt & Tsividis 1987 [5] (Columbia)	Sampled-signal z-domain filters	✔	✔	✔				
Sivilotti 1988 [19] (Caltech)	Subthreshold		✔					
Lee & Gulak 1990 [16] (Toronto)	Subthreshold	✔	✔	✔				
Lee & Gulak 1992 [8] (Toronto)	Linear	✔	✔	✔	✔			✔ [66]
Pierzchala & Perkowski 1994 [10] (Portland State)	Linear	✔						
Chang et al. 1994 [9] (Nottingham)	Linear	✔						
Chang et al. 1996 [7] (Nottingham)	Switched-current	✔						
Premont et al. 1996 [11] (Cimirly Lyon)	Current conveyor	✔	✔	✔				✔
Embabi et al. 1996 [12] (Texas A&M)	Current integrator	✔	✔	✔				
Pilkington–Motorola 1993 [21]	Switched-capacitor	✔			✔	✔		✔
Kawasaki Steel 1994 [81]	Linear					✔		
IMP 1994 [77]	Switched-capacitor	✔	✔	✔	✔		✔	
Zetex 1996 [78]	Linear	✔	✔	✔	✔		✔	

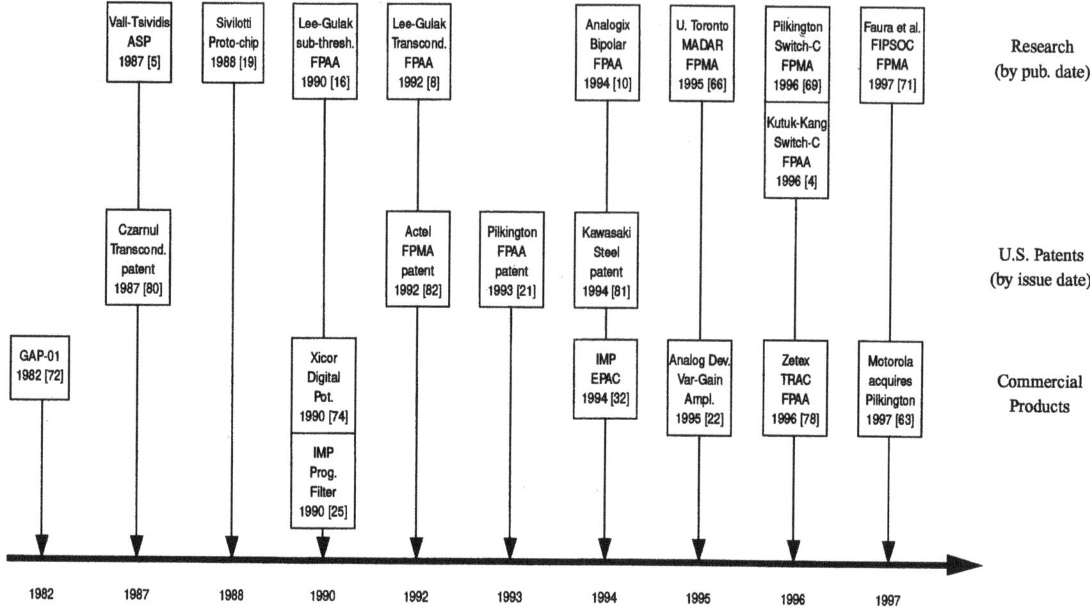

Fig. 17. Timeline marking important milestones in the field of FPAAs/FPMAs.

time-to-market pressures now universally exerted on product developers. The acquisition of PMeL by Motorola, and its announcement of plans to release FPAA and FPMA IC products further indicates the commercial potential of this field.

Several developments can be expected in the future. Present switched-capacitor techniques show good promise for applications below 1 MHz; this and other sampled data techniques can be expected to yield programmable analog and mixed-signal applications in lower bandwidth applications such as measurement, monitoring and control. In the longer term, it is conceivable that switched-capacitor techniques might support applications up to 10 MHz. The use of stochastic signal processing techniques on a Pulse Density Modulated (PDM) representation of analog signals might lead to some interesting implementations for low-frequency applications in measurement and monitoring. As demonstrated in [86], a PDM bit-stream produced using Sigma-Delta modulation analog to digital conversion can be processed using stochastic or deterministic methods that require simple digital hardware, and reconfigurability can be achieved using a digital FPGA.

For higher frequency applications, for example processing of video bandwidth signals, existing continuous-time techniques will need to be improved by the development of high-performance reconfigurable analog circuit designs, like the reconfigurable opamp/differential difference amplifier in [87]. Improved architectures and implementations of FPAA interconnect will be especially important in this respect.

Future developments in both analog and digital integrated circuit memory will improve the density and performance of FPAAs. Circuit configuration and component parameter storage occupies a significant percentage of FPAA area, and the accuracy of circuit component values requires analog memory or precision sample and hold amplifiers with accuracies greater than ten bits. The area of ferroelectric memory shows promise for non-volatile memory at densities competitive with DRAM, and requires simpler IC processing steps and programming techniques than flash memory. Current research in the area [88] is addressing multiple valued ferroelectric memory.

A tighter coupling of FPAAs to digital structures is already evident in the evolution of the IMP EPAC devices [32]. This can be expected to continue to address interfacing of these devices to microcontrollers and monitoring systems, as well as for providing Built In Self Test (BIST) and calibration features to analog circuits. Adaptive operation, whereby a device reconfigures itself on the fly as different functions are required, might also be attempted as has been done with digital FPGAs [89,90]. Field-programmable mixed-signal systems might also be implemented by coupling FPGA and FPAA ICs at the Multi-Chip Module or Printed Circuit Board levels. Research is necessary in the area of mixed-signal arrays, to determine circuit resources for FPMAs, particularly for the interface between analog and digital converters. Reconfigurable data converters [66,91] will be important in this respect.

Specifications for current FPAA designs have been mostly in the form of the publication of performance numbers for CABs or programmable circuit elements. Some works have addressed the issue of interconnect by looking at parasitic effects, but as yet no standards have emerged for succinct methods of describing pin-to-pin performance. An exception to this has been restricted architecture programmable circuits such as the IMP EPAC devices [32], where all possible circuit configurations can be predicted and worst case specifications determined. A useful design aid would be a software program that understands the parameters and interactions of CABs and interconnect, and can predict performance for the embedding of a circuit on a target FPAA from the configuration data.

Acknowledgments

The authors thank NSERC, Micronet and ITRC for financial support. The helpful comments of Edward Lee and Vincent Gaudet are gratefully acknowledged. For additional information on FPAAs and FPMAs consult the web page http://www.eecg.toronto.edu/ ~ vgaudet/fpaa.html.

References

1. M. Ismail and S. Bibyk, ''CAD Latches Onto New Techniques for Analog ICs.'' *IEEE Circuits and Devices Magazine*, September 1991, pp. 11–17.
2. W. J. McClean, ed., ''Status 1996, A Report on the Integrated Circuit Industry.'' Integrated Circuit Engineering Corp., 1996.
3. A. Bratt and I. Macbeth, ''Design and Implementation of a FPAA.'', *ACM/SIGDA FPGA'96*, Monterey, Ca., Feb. 11–13, 1996, pp. 88–93.
4. H. Kutuk and S. Kang, ''A Field-Programmable Analog Array (FPAA) Using Switched-Capacitor Techniques.'' *IEEE ISCAS 1996* 3, pp. 41–44.
5. D. Vallancourt and Y. P. Tsividis, ''Timing-Controlled Switched Analog Filters with Full Digital Programmability.'' *IEEE ISCAS 1987*, pp. 329–333.
6. Adaptive Logic, AL220 Analog Micro Controller preliminary data sheet, WWW site: http://www.adaptivelogic.com
7. S. T. Chang, B. R. Hayes-Gill, and C. J. Paull, ''Multi-Function Block for a Switched Current Field Programmable Analog Array.'' *Midwest Symposium on Circuits and Systems*, Ames, Iowa, August 18–21 1996.
8. E. Lee and P. G. Gulak, ''Field-Programmable Analogue Array based on MOSFET Transconductors.'' *Electronic Letters* 28(1), pp. 28–29, January 2, 1992.
9. S. Chang, B. Hayes-Gill, and C. Paull, ''Implementation of a Multi-Function Signal Detection Block for a Field-Programmable Analogue Array.'' *Fifth Eurochip Workshop on VLSI Design Training*, Oct. 17–19, 1994, Dresden, Germany, pp. 226–231.
10. E. Pierzchala, M. Perkowski, Paul Van Halen, and Rolf Schaumann, ''Current-Mode Amplifier-Integrator for a Field-Programmable Analog Array.'' *ISSCC Digest of Technical Papers*, Feb. 1995, pp. 196–197.
11. C. Premont, R. Grisel, N. Abouchi, and J Chante, ''Current-Conveyor Based Field Programmable Analog Array.'' *Midwest Symposium on Circuits and Systems*, Ames, Iowa, August 18–21 1996.
12. S. H. K. Embabi, X. Quan, N. Oki, A. Manjrekar, and E. Sánchez-Sinencio, ''A Field Programmable Analog Signal Processing Array.'' *Midwest Symposium on Circuits and Systems*, Ames, Iowa, August 18–21 1996.
13. H.W. Klein, ''The EPAC Architecture: An Expert Cell Approach to Field Programmable Analog Devices.'' *ACM/SIGDA FPGA'96*, Monterey, CA, Feb. 11–13, 1996, pp. 94–98.
14. C. Toumazou and J. Lidgely, ''Universal Current-Mode Analogue Amplifiers.'' In *Analogue IC Design: the Current-Mode Approach*, Peter Peregrinus Ltd., London, UK, 1990, pp. 127–180.
15. G. Roberts, ''Generalization and Application of the Intermediate Function Technique.'' *PhD. Dissertation*, University of Toronto, 1989.
16. E. Lee and P. G. Gulak, ''Prototype Design of a Field-Programmable Analog Array.'' in *Proceedings of CCVLSI 1990*, Ottawa, Canada, October 21–23, 1990.
17. C. Toumazou, J. B. Hughes, and N. C. Battersby eds., *Switched-Currents an Analogue Technique for Digital Technology*, Peter Peregrinus: UK, 1993.
18. E. Lee, ''Field-Programmable Analog Arrays Based on MOS Transconductors.'' *PhD. dissertation*, University of Toronto, 1995.
19. M. Sivilotti, ''A Dynamically Configurable Architecture for Prototyping Analog Circuits.'' *MIT VLSI Conference*, 1988, pp. 237–258.
20. W. A. Fisher, R. J. Fujimoto, and R. C. Simpson, ''A Programmable Analog Neural Network Processor.'' *IEEE Transactions on Neural Networks* 2(2), pp. 222–228, March 1991.
21. K. Austin, ''Integrated Circuit for Analogue System.'' U.S. Patent 5,196,740, Pilkington Micro-Electronics, March 23, 1993.

22. Analog Devices, "Low-Noise, 90MHz Variable-Gain Amplifier." AD603 Data Sheet, 1995.
23. Burr-Brown, "Digitally-Controlled Programmable-Gain Instrumentation Amplifier." PGA200/201 Data Sheet, 1987.
24. Burr-Brown, "Universal Active Filter." UAF42, Data Sheet, 1990.
25. F. Goodenough, "Voltage-Tunable Linear Filters Move onto a Chip." *Electronic Design*, February 8, 1990.
26. F. J. Kub, I. A. Mack, K. K. Moon, C. T. Yao, and J. A. Modolo, "Programmable Analog Synapses for Microelectronic Neural Networks using a Hybrid Digital-Analog Approach." presented at the *IEEE International Conference on Neural Networks*, San Diego, CA, July 24–27, 1988.
27. F. J. Kub, K. K. Moon, I. A. Mack, and F. M. Long, "Programmable Analog Vector-Matrix Multipliers." *IEEE Journal of Solid State Circuits* 25(1), pp. 207–214, February 1990.
28. S. Satyanarayana, Y. P. Tsividis, and H. P. Graf, "A Reconfigurable VLSI Neural Network." *IEEE Journal of Solid State Circuits* 27(1), pp. 67–81, January 1992.
29. E. Lee and P. G. Gulak, "Error Correction Technique for Multivalued MOS Memory." *Electronics Letters* 27(15), pp. 1321–1323, July 18, 1991.
30. E. Lee and P. G. Gulak, "Current-Mode Multivalued Dynamic MOS Memory with Error Correction." *Electronics Letters* 28(11), pp. 1067–1069, May 21, 1992.
31. Pilkington Micro-electronics Limited (PMeL) WWW site: http://www.pmel.com.
32. IMP Inc., "Programmable Analog Signal-Conditioning Circuit." IMP50E10 and IMP50E30 Data sheets, 1994.
33. T. Roiné, "The FPAA Interface, manual and report" Technical Report, University of Toronto, December 1995.
34. S. Lecler, "The FPAA Programming Tool, manual and report." Technical Report, University of Toronto, 1994.
35. R. A. Saleh, D. L. Rhodes, E. Christen, and B. A. A. Antao, "Analog Hardware Description Languages." *IEEE Custom Integrated Circuits Conference 1994*, pp. 349–356.
36. WWW site for the IEEE VHDL-A standard, http://vhdl.org/vi/analog/.
37. J. Bergé, O. Levia, and J. Rouillard eds., *Modeling in Analog Design*, Kluwer Academic Publishers: The Netherlands, 1995.
38. A. F. Arbel, "Current Operated Nucleonic Modules." *Nuclear Instruments and Methods* 32, pp. 341–346, 1965.
39. R. Joetten, T Web, J. Wolters, H. Ring, and B. Bjoernsson, "A New Real-Time Simulator for Power System Studies." *IEEE Transactions on Power Apparatus and Systems* PAS-104(9), pp. 2604–2611, September 1985.
40. F. Goodenough, "Analog Arrays Rival Digital Designs in Facility, Performance." *Electronic Design* pp. 86–96, June 13, 1985.
41. G. Gianella, "Array IC Presents New Ways to Customize Analog Circuits Without Wasting Silicon." *Electronic Design* pp. 171–178, May 1, 1986.
42. L. B. Wheaton, "Universal Cell for Bipolar NPN and PNP Transistors and Resistive Elements." U.S. Patent 5,021,856, Plessey Overseas Limited, June 4, 1991.
43. K. Negus, R. Koupal, D. Millicker, and C. Snapp, "Silicon Bipolar Mixed-Signal Parameterized-Cell Array for Wireless Applications to 4GHz.", *ISSCC Digest of Technical Papers*, San Francisco, CA, pp. 230–231, 1992.
44. P. P. Duchene, M. J. Declerq, B. Goffart, and M. Novak, "Analog Circuit Implementation on CMOS Semi-Custom Arrays." *IEEE Journal of Solid-State Circuits* 28(7), pp. 872–874, July 1993.
45. S. Shinbara, "Analog Standard Cell." U.S. Patent 5,302,864, Kabushiki Kaisha Toshiba, April 12, 1994.
46. S. L. Wong, S. Venkitasubrahmanian, M. J. Kim, and J. C. Young, "Design of a 60-V 10-A Intelligent Power Switch Using Standard Cells." *IEEE Journal of Solid State Circuits* 27(3), pp. 429–432, March 1993.
47. M. Soyuer, J. N. Burghartz, K. A. Jenkins, H. A. Ainspan, and F. J. Canora, "RF and Microwave Building Blocks in a Standard BiCMOS Technology." *IEEE ISCAS 1996*, 4, pp. 89–92.
48. R. C. Chang, B. J. Sheu, J. Choi, and D. C-H Chen, "Programmable Weight Building Blocks for Analog VLSI Neural Network Processors." *Analog Integrated Circuits and Signal Processing* 9(3), pp. 215–230, April 1996.
49. J. Van der Spiegel, P. Mueller, D. Blackman, P. Chance, C. Donham, R. Etienne-Cummings, and Peter Kinget, "An Analog Neural Computer with Modular Architecture for Real-Time Dynamic Computations." *IEEE Journal of Solid State Circuits* 27(1), pp. 82–92, January 1992.
50. S. Eberhardt, T. Duong, and A. Thakoor, "Design of Parallel Hardware Neural Network Systems from Custom Analog VLSI 'Building Block' Chips." *IEEE INNS Intl. Joint Conference on Neural Networks* 2, pp. 183–190, June 1989.
51. B. J. Sheu, "VLSI Neurocomputing with Analog Programmable Chips and Digital Systolic Array Chips." *Proc. IEEE ISCAS 1991*, pp. 1267–1270.
52. E. Lee and G. Gulak, "A CMOS Field-Programmable Analog Array." *ISSCC Digest of Technical Papers*, Feb. 1991, pp. 186–188.
53. E. Lee and G. Gulak, "A Transconductor-Based Field-Programmable Analog Array." *ISSCC Digest of Technical Papers*, Feb. 1995, pp. 198–199.
54. L. O. Chua and L. Yang, "Cellular Neural Networks: Theory and Applications." *IEEE Transactions on Circuits and Systems* 35(10), pp. 1257–1290, October 1988.
55. P. Kinget and M. S. J. Steyaert, "A Programmable Analog Cellular Neural Network CMOS Chip for High Speed Image Processing." *IEEE Journal of Solid-State Circuits* 30(3), March 1995.
56. P. Kinget and M. Steyaert, "An Analog Parallel Array Processor for Real-Time Sensor Signal Conditioning." *IEEE ISSSCC Digest of Technical Papers*, Feb. 1996, pp. 92–93.
57. T. Roska and L.O. Chua, "The CNN Universal Machine: An Analogic Array Computer." *IEEE Transactions on Circuits and Systems II* 40(3), pp. 163–173, March 1993.
58. H. C. Anderson, "Complex Signal Transformation Using a Resistive Network." U.S. Patent 5,058,049, Motorola Inc., October 15, 1991.
59. E. Lee and G. Gulak, "A CMOS Field-Programmable Analog Array." *IEEE Journal of Solid-State Circuits* 26(12), pp. 1860–1867, December 1991.
60. D. Vallancourt and Y. P. Tsividis, "Timing-Controlled Fully Programmable Analog Signal Processors Using Switched Continuous-Time Filters." *IEEE Transactions on Circuits and Systems* 35(8), pp. 947–954, August 1988.

61. Pilkington Micro-Electronics, ''Programmable Switched Capacitor Circuit.'' UK Patent Application GB 2,275,144A, 17/08/1994.
62. P. Clarke, ''Analog-cell IC takes reconfigurable tack.'' *Electronic Engineering Times*, October 7, 1996, p. 1.
63. P. Clarke, ''Moto buys Pilkington, gets programmable logic.'' *Electronic Engineering Times*, March 24, 1997, p. 18.
64. Z. Czarnul, ''Novel MOS Resistive Circuit for Synthesis of Fully-Integrated Continuous-Time Filters.'' *IEEE Trans. Circuits Syst.*, CAS-33, pp. 718–722, July 1986.
65. E. Pierzchala, M. A. Perkowski, and S. Grygiel, '' A Field-Programmable Analog Array for Continuous, Fuzzy, and Multi-Valued Logic Applications.'' *IEEE ISMVL*, Boston, Mass., May 1994, pp. 148–155.
66. P. Chow, P. Chow, and P.G. Gulak, ''A Field-Programmable Mixed Analog-Digital Array.'' *ACM/SIGDA FPGA'95*, Monterey, CA, Feb. 12–14, 1995, pp. 104–109.
67. P. Chow, S. O. Seo, K. Chung, G. Paez, and J. Rose, ''A High-Speed FPGA Using Programmable Mini-tiles.'' *Proc. 1993 Symposium of Integrated Systems*, March 1993.
68. K. Bult and G. Geelen, ''An Inherently Linear and Compact MOST-Only Current-Division Technique.'' *IEEE JSSC* 27(12), pp. 1730–1735, December 1992.
69. C. Zhang, A. Bratt, and I. Macbeth, ''A New Field Programmable Mixed Signal Array and its Application.'' *4th Canadian Workshop on Field-Programmable Devices*, May 13–14, 1996, Toronto, Canada.
70. P. Clarke and R. Wilson, ''Pilkington Details its Analog-Array Effort.'' *Electronic Engineering Times*, September 25, 1995, p. 4.
71. J Faura, C. Horton, P. van Duong, J Madrenas, M. A. Aguirre, and J. M. Insenser, ''A Novel Mixed Signal Programmable Device with On-chip Microprocessor.'' *IEEE Custom Integrated Circuits Conference 1997.*
72. Precision Monolithics Inc., ''Analog Signal Processing Subsystem.'' GAP-01 Data Sheet, 1982.
73. P. Henneuse, ''Low Glitch Current Switch.'' U.S. Patent 4,285,051, Precision Monolithics Inc., August 18, 1981.
74. Xicor, ''Nonvolatile Digital Potentiometer.'' X9C102 Data Sheet, 1994.
75. Analog Devices, ''2-/4-Channel Digital Potentiometer.'' AD8402/8403 Data Sheet, 1995, WWW site: http://www.analog.com
76. Texas Instruments, TLC3203X/4X Data Sheet, 1994, WWW site: http://www.ti.com.
77. F. Goodenough, ''Analog Counterparts of FPGAs Ease System Design.'' *Electronic Design*, Oct. 14, 1994, pp. 63–73.
78. Zetex Semiconductors WWW site: http://www.zetex.com.
79. R. W. Harris and H. T. Lee, ''Digitally Controlled Variable Conductance.'' U.S. Patent 4,009,400, Lockheed Missiles and Space Company Inc., February 22, 1977.
80. Z. Czarnul, ''Semiconductive MOS Resistance Network.'' U.S. Patent 4,710,726, Columbia University, December 1, 1987.
81. N. Sako, ''Integrated Circuit and Gate Array.'' U.S. Patent 5,298,806, Kawasaki Steel Corp., March 29, 1994.
82. K. A. El-Ayat, ''Mixed Mode Analog/Digital Programmable Interconnect Architecture.'' U.S. Patent 5,107,146, Actel Corporation, April 21, 1992.
83. O. Agrawal, M. Wright, and S. Sidman, ''Programmable Logic Device Incorporating Voltage Comparator.'' U.S. Patent 5,153,462, Advanced Micro Devices, Oct. 6, 1992.
84. O. Agrawal and M. Wright, ''Programmable Logic Device Incorporating Digital-to-Analog Converter.'' U.S. Patent 5,191,242, Advanced Micro Devices, March 2, 1993.
85. R. A. Baumgartner and E. C. Herleikson, ''Signal Processing Circuits with Digital Programmability.'' U.S. Patent 5,337,230, Hewlett-Packard Company, August 9, 1994.
86. R. Ananth, ''Signal Processing Structures for Power-Supervisory and Low-Frequency Applications.'' MASc. Thesis, University of Toronto, 1992.
87. S. R. Zarabadi, F. Larsen, and M. Ismail, ''A Reconfigurable Op-Amp/DDA CMOS Amplifier Architecture.'' *IEEE Trans. Ccts & Sys.-I: Analog and Digital Sig. Proc.* 39(6), June 1992.
88. A. Sheikholeslami, P. G. Gulak, and T. Hanyu, ''A Multiple-Valued Ferrolectric ContentAddressable Memory.'' *IEEE ISMVL*, Santiago de Compostela, Spain, May 1996.
89. M. J. Wirthlin and B. L. Hutchings, ''Sequencing Run-Time Reconfigured Hardware with Software.'' *ACM/SIGDA FPGA'96*, Monterey, Ca., Feb. 11–13, 1996, pp. 122–128.
90. Xilinx Inc., ''XC6200 Field-Programmable Gate Arrays.'' Data Sheet, 1996, WWW site: http://www.xilinx.com
91. E. K. F. Lee, ''Reconfigurable Pipelined Data Converter Architecture.'' *Midwest Symposium on Circuits and Systems*, Ames, Iowa, August 18–21 1996.

Dean D'Mello is an R&D Design Engineer in the Consumer Products Division of LSI Logic Inc. in Toronto, Canada.

His Research Interests include Field-Programmable Integrated Circuits and Computer-aided design of analog and digital ICs.

He was awarded a Postgraduate scholarship by the Natural Sciences and Engineering Research Council of Canada, and received a MASc. in Electrical and Computer Engineering from the University of Toronto in 1996, with a research thesis on layout synthesis of Field-Programmable Analog Array cores.

From 1991 to 1994 he was an Associate Engineer in the Power Products Division of IBM Canada Ltd. From 1996 to 1997 he was a Senior Associate Engineer in the Mixed-Signal & Power ASIC Development Group at Celestica Inc. in Toronto.

In 1991, he graduated with a BASc. in Electrical Engineering from the University of Toronto, and received the Centennial Thesis Award for a thesis entitled ''Spreadsheet Models of Hopfield Neural Networks.''

Glen Gulak is a professor in the Department of Electrical and Computer Engineering at the University of Toronto.

He is a senior member of the IEEE and a registered professional engineer in the province of Ontario.

His research interests are in the areas of circuits, algorithms and VLSI architectures for digital communications and signal processing applications. He has received several teaching awards for undergraduate courses taught in both the Department of Computer Science and the Department of Electrical and Computer Engineering at the University of Toronto.

Dr. Gulak received his Ph.D. from the University of Manitoba while holding a Natural Sciences and Engineering Research Council of Canada Postgraduate Scholarship. From 1985 to 1988 he was a Research Associate in the Information Systems Laboratory and the Computer Systems Laboratory at Stanford University. He has served on the ISSCC Signal Processing Technical Subcommittee since 1990 and currently serves as Program Secretary for ISSCC.

Analog Integrated Circuits and Signal Processing, 17, 35–50 (1998)

A Novel Switched-Capacitor Based Field-Programmable Analog Array Architecture

EDWARD K. F. LEE AND WAI L. HUI

Department of Electrical and Computer Engineering, Iowa State University, Ames, IA 50011

Received July 15, 1996; Accepted April 24, 1997

Abstract. A novel field-programmable analog array (FPAA) architecture based on switched-capacitor techniques is proposed. Each configurable analog block (CAB) in the proposed architecture is an opamp with feedback switches which are controlled by configuration bits. Interconnection networks are used to connect programmable capacitor arrays (PCAs) and the CABs. The routing switches in the interconnection networks not only function as interconnection elements but also switches for the charge transfer required in switched-capacitor circuits. This scheme minimizes the number of connecting switches between CABs and PCAs, thereby, it reduces the settling time of the resultant SC circuits and thus achieving high speed operation. The architecture is highly flexible and provides for the implementation of various A/D and D/A converters when the FPAA is connected with external digital circuits or field-programmable gate arrays (FPGAs).

Key Words: FPAA, programmable analog array

1. Introduction

Rapid-prototyping techniques for analog circuits analogous to the use of field-programmable gate arrays (FPGAs) for prototyping digital circuits have been recently introduced [1–6]. These techniques are based on the use of field-programmable analog arrays (FPAAs) which are defined as integrated circuits that can be programmed by the user to implement analog circuits using configurable analog blocks (CABs) and programmable interconnections as shown in Fig. 1. A given circuit is implemented by loading configuration bits into the on-chip registers which determine the function of the CABs and the connections made by the interconnection network. Advantages of using the FPAA not only include instant prototyping but also reconfigurability, CAD compatibility, parameter programmability and testability [2].

Different FPAA design techniques have been proposed including operating MOSFETs in the subthreshold region [1,2,7], the use of MOSFET transconductors [3,4], and the use of bipolar current-mode techniques [5,6]. Recently, switched-capacitor (SC) techniques for implementing FPAAs have been explored in research laboratories [8] and have found application in commercial products [9–12]. These products support general purpose signal conditioning for medical, industrial or other control and instrumentation systems. Unfortunately, the flexibility of these designs is limited and large area is required to implement complex circuits such as high order SC filters with various structures, and different types of A/D and D/A converters. As a result, direct integration with FPGAs for realizing field-programmable mixed analog and digital arrays (FPMAs) [12,13] may be difficult due to large area requirement on the FPAA. Furthermore, the large signal bandwidth is limited in the range of kHz. In contrast to other implementation approaches, SC techniques are sampled-data in nature and require both anti-aliasing and reconstruction filters. Circuit parameter programming is usually achieved by changing the capacitance values of the programmable capacitor arrays (PCAs), an approach that requires large area for high resolution. Despite these draw backs, SC techniques can provide high accuracy and do not require the multi-valued memories for storing circuit parameters that are usually required in FPAAs implemented using other techniques [2,4]. In this paper, a new FPAA architecture based on the SC technique that achieves high flexibility, area efficient and high speed operation is proposed. Specialized circuits such as sigma-delta modulators, A/D and D/A converters can be effectively implemented in the proposed architecture.

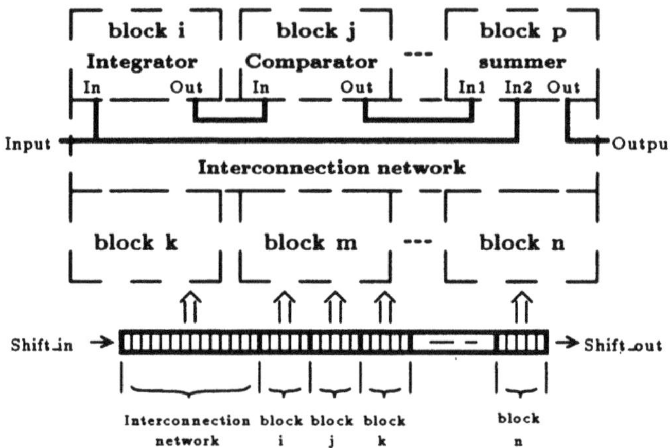

Fig. 1. A conceptual field-programmable analog array.

Therefore, the architecture is very suitable for direct integration with FPGAs for constructing FPMAs.

In Section 2, general design strategies for the SC-based FPAA with different granularity levels are presented. Based on the granularity level selected, a flexible FPAA architecture is introduced in Section 3. Implementation of various SC circuits including A/D and D/A converters are illustrated in Section 4. Conclusions of this paper are presented in Section 5.

2. Granularity Levels

The main circuit elements in SC circuits are switches, opamps and capacitors. Depending on the granularity level (how these components are grouped together), different SC-based FPAA architectures can be obtained. In general, one can design a FPAA under two different granularity levels—the macro-block level or building-block level. For FPAAs based on the macro-block level, a configurable analog block (CAB) will consist of an opamp, switches, programmable capacitor arrays (PCAs) and a control unit that controls the functions performed by the CAB. Circuit functions such as integration, summing, sample-and-hold, programmable gain amplification, etc., can be realized by programming the connections between the opamp, switches and PCAs within the CAB. Most SC based FPAA architectures [8,10,12] have their CABs implemented at this granularity level. Fig. 2 shows a possible CAB design that can be programmed as an integrator or a summing amplifier. To implement an analog system, the CABs are programmed to realize different functions and connections between the CABs are made with an interconnection network. If the opamps have a very high unity gain frequency[1] and the capacitance associated with the connection wire is negligible, the maximum clock rate f_{max} for 0.1% accuracy is approximately equal to $1/20RC_{min}$ [14] where C_{min} is the minimum capacitance value programmed to the PCAs and R is the summation of the resistance of the routing switches in each routing path with the resistance of the charge transfer

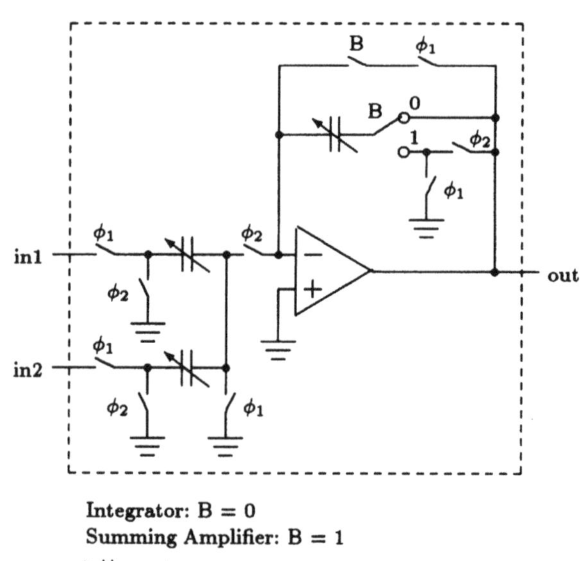

Fig. 2. A macro-block level CAB that can be used as an integrator or a summing amplifier.

switches. Since several routing switches are required to make a connection between two CABs, f_{max} will be limited to a few MHz.

At the building-block level, the CAB will consist of either an opamp or a PCA. Analogous to the ideal proposed in [3] where routing switches in the interconnection network are utilized as circuit components, the routing switches in a building-block level FPAA can be used as the switches required for charge transfer in a normal SC circuit. The concept is illustrated in Fig. 3. Since the number of switches between the opamps and capacitors are decreased, the resistance associated with the switches will be reduced and the settling time will be decreased. A maximum clock rate of a few ten's of MHz can be achievable using switches with reasonable size.

When the macro-block level approach and the building-block level approach are compared, the later approach is more flexible and more area efficient. As an illustration, if the CAB shown in Fig. 2 is used for implementing a circuit that performs a weighted sum of three different signals, two CABs are required and the combined area of the two CABs are approximately equivalent to the area of two opamps and six PCAs. In a normal SC design, only one opamp and four capacitors are needed. For the building-block level approach, four PCAs and one opamp are needed resulting in a more area efficient realization. Although one may argue that the area of the interconnection network for the building block level may be larger than that of the macro block level, this area is less important when it is compared with the area of the PCAs which grows exponentially as the resolution of the PCAs increases. If a circuit requires a capacitor which is larger than the value that can be accommodated by a single PCA, multiple PCAs can be connected in parallel when using the building-block level FPAA. This flexibility allows the PCAs in building-block level FPAAs to have finer resolution than the PCAs in the macro-block level FPAAs. An FPAA architecture based on the building-block level approach which supports higher clock rates and provides more flexibility and better area efficiency is introduced in the following section.

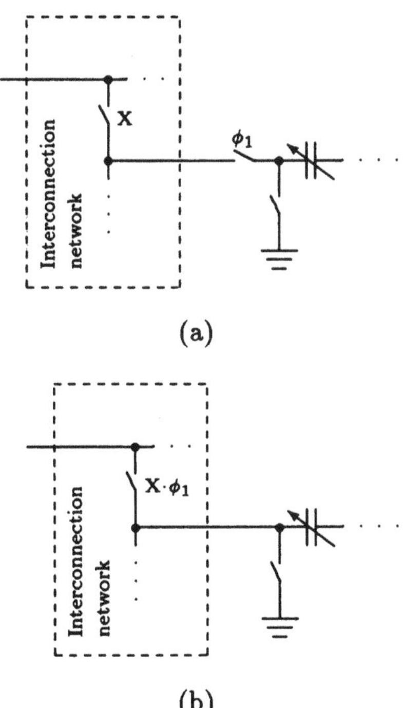

Fig. 3. (a) Separated switches for charge transfer and interconnection and (b) routing switches functioned as switches for charge transfer.

3. Proposed Architecture

Observation of different SC circuits [14,15] show that most SC circuits require an average 2—3.5 of capacitors (PCAs) per opamp. To achieve high utilization, the ratio between PCAs and opamps in a FPAA will be selected based on the average value. Observations show that each switch in a SC circuit usually has one end connected to the input/output terminals of an opamp and the other end connected to a capacitor. There are a few exceptional cases such as the switch for connecting the opamp into a unity-gain configuration. Observations also show that the opamps and capacitors are usually connected locally. Therefore, only a few long interconnects are need in an FPAA. Further observations show that several clock phases are required in some SC circuits. For sigma-delta modulators and some A/D converters implemented with SC techniques, some of the opamp outputs are required to control the on/off state of the switches.

A flexible SC based FPAA that accommodates these observations and that can achieve a high clock rate is proposed. The general architecture with PCA-to-opamp ratio equal to two is shown in Fig. 4. Note that each line in the figure represents a pair of fully

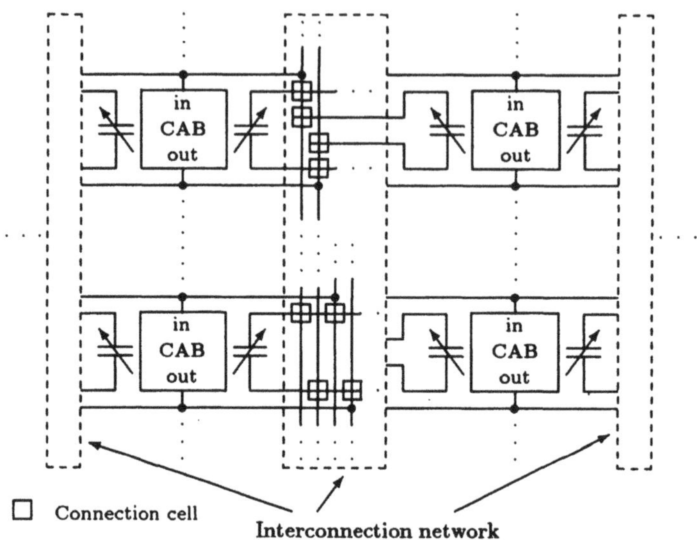

Fig. 4. Proposed SC based FPAA architecture.

differential wires. The CAB consists of an opamp and two feedback switches as shown in Fig. 5. The feedback switches are controlled by selected global clock signals that depend on the configuration bits loaded into the CAB. The opamp is assumed to be capable of reconfiguring into a comparator. The design of the PCA is shown in Fig. 6[2] where the capacitance of the PCA is determined by the digital inputs (b_s to b_n) which is loaded into the PCA through a data bus after the PCA is selected by loading the corresponding address to an address bus. Both ends of the PCA can be shorted to ground according to the selected clock signals. The main feature of this architecture is that the number of switches for making a connection between a CAB and a capacitor is reduced due to the utilization of the connection switches in the interconnection networks as the switches required in a normal SC circuit. The PCAs and the CABs in one column can be connected to the PCAs and the CABs in the adjacent columns via the connection cells as illustrated in Fig. 4. The details of the connection cells are shown in Fig. 7. The gates of the MOSFET switches can be connected to the different clock signals, 1 (V_{dd}) or 0 (V_{ss}). The switches are either turned on/off or controlled by the clock signals which are assumed to be distributed to all connection cells. Since there is always only one switch between a CAB and a capacitor, the settling time is approximately the same as a normal SC circuit if the parasitic capacitances of the interconnection wires are neglected.[3] In the output connection cell (Fig. 7b), the connections between the auxiliary input (connected to all the output connection cells in the same column) and the PCA terminals are controlled by clock signals P1 or P2, and the CAB outputs, O and O!. This feature is useful for implementing sigma-delta modulators and A/D converters as discussed in Section 4.

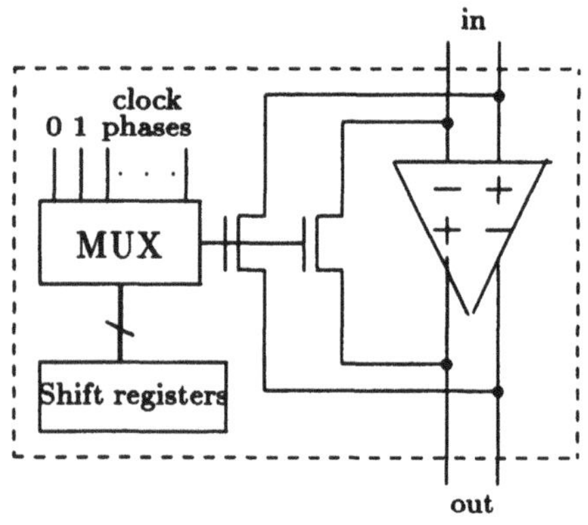

Fig. 5. CAB design for the proposed FPAA.

To demonstrate that the proposed architecture can have higher clock rates than that of the macro-block level FPAAs, the circuits shown in Fig. 8 were simulated based on a 2 μm CMOS process. All the transistors are assumed to have minimum size. A capacitor of 0.1 pF was added to each node. The opamp was assumed to have a unity gain frequency at

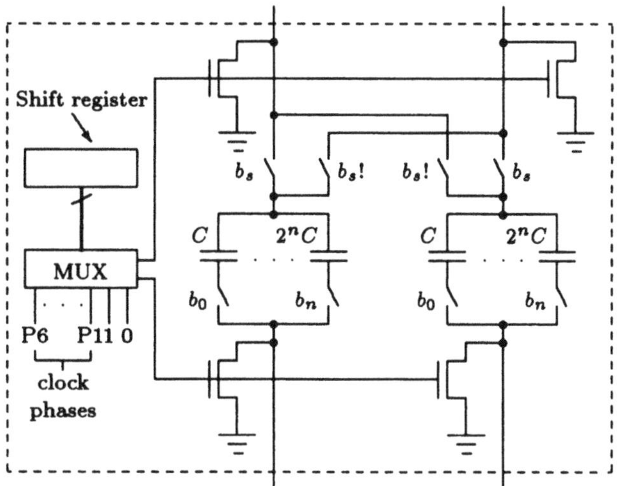

Fig. 6. Programmable capacitor array (PCA).

50 MHz, and the capacitor C_1 and C_2 are set to 1 pF and 2 pF, respectively. V_i is assumed to be constant. Fig. 9 shows the opamp output voltages when P1 turns high. The macro-block level FPAA requires 248.4 ns to settle for 0.1 % accuracy and the proposed architecture only requires 130.9 ns to settle. Since the proposed architecture utilizes the routing switches as the charge transfer switches in the SC circuits, the proposed architecture has a shorter settling time. Clock frequency in the range of a few tens of MHz can be achievable if the width-to-length ratio of the routing switches in the proposed architecture are larger than minimum size.

The area of the interconnection network is dependent on the wire spaces W_w and the wire length of the CAB input and output terminals R which is in terms of a CAB or a PCA pitch. These two parameters determines the width of the interconnection network while the number of PCAs or CABs in a column determines the height of the interconnection network. Since the opamps and the capacitors in SC circuits are usually locally connected, an R value of three is found to be long enough for implementing most SC circuits. Assuming the width of two wire spaces is equal to the width of one connection cell, W_{CC}, the width of each interconnection network for a single column architecture (Fig. 10) is equal to $2RW_w$ which is equal to $3W_{CC}$. For the multi-column architecture (Fig. 4), if we assume W_{CC} is still equal to $2W_w$, the width of each interconnection network is equal to $6W_{CC}$ independent upon the number of PCAs or CABs in a column. Therefore, the proposed architecture is area efficient since it only exhibits a linear growth of area for increasing CABs and PCAs.

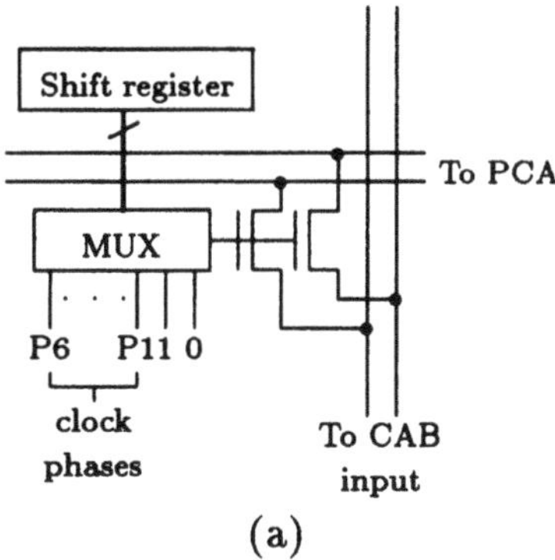

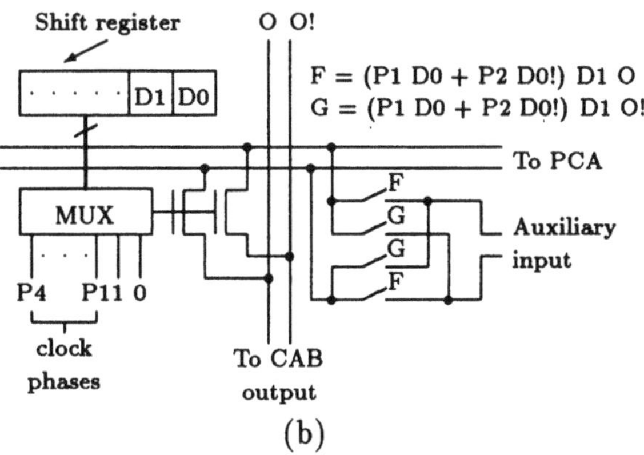

Fig. 7. (a) Input connection cell and (b) output connection cell.

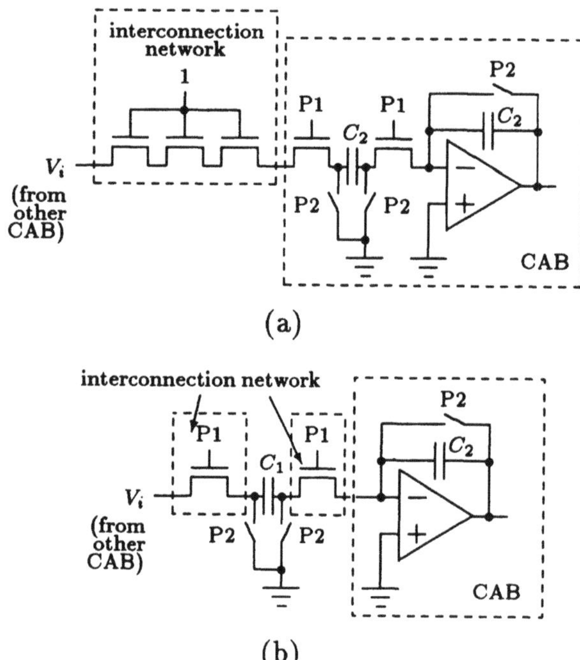

Fig. 8. (a) Model for macro-block level FPAAs and (b) model for the proposed FPAA.

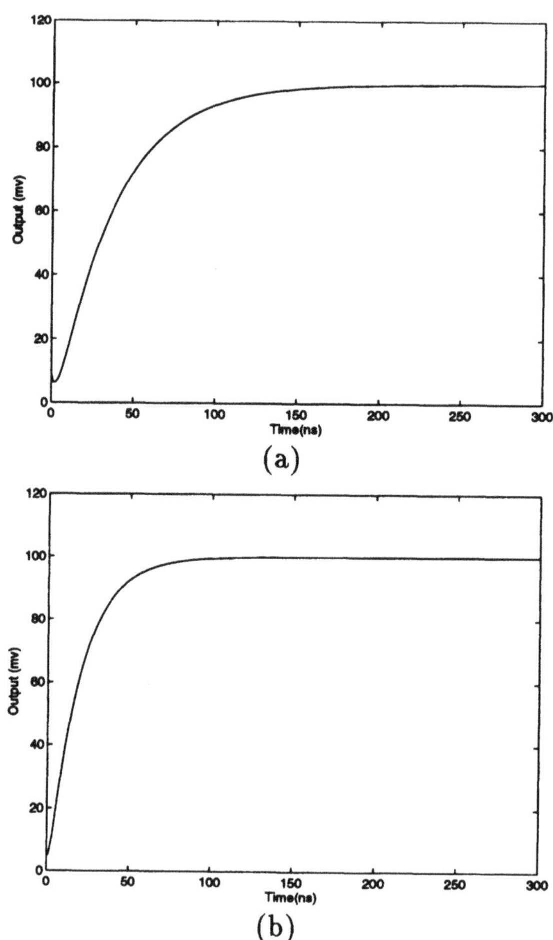

Fig. 9. Simulated transient behaviors for (a) the macro-block level FPAAs and (b) the proposed FPAA.

For actual implementation, interconnection networks with slightly wider width will be used to allow some of the CAB input/output terminals to have length more than three pitches. Since most of the CAB input/output terminals have short lengths, the parasitic capacitance associated with the wires will be less. Hence, the speed of operation can be increased and the cross-talk between wires will be reduced. Although increasing the number of clock signals for controlling the switches will increase the width of the connection cell, the interconnection network area will still be less important than the area of the PCAs and CABs, especially when a fully differential architecture with high resolution PCAs is used.

4. Circuit Examples

The proposed architecture is highly flexible and can be used to implement numerous different SC circuits. A few circuit examples are discussed below. The embedding of these examples are demonstrated using an architecture that has only one column as shown in Fig. 10. In the following sections, a label inside a connection cell indicates the clock signal controlling the MOSFET switches. Similarly, the labels indicated at the ends of a PCA represent the clock signals that control the switches between the PCA and ground (referred to Fig. 6). If an output connection cell with a label in the center is highlighted, this indicates that the output connection cell is controlled by the opamp output and the labelled clock signal. When a label appears at the left of a CAB, the feedback switches in the CAB are controlled by the labelled clock signal. If there is no label on the connection cells, CABs or PCAs, the switches in these components are off. Although the circuits discussed in the following sections are drawn as single-ended, the implementation in the FPAA will be fully differential.

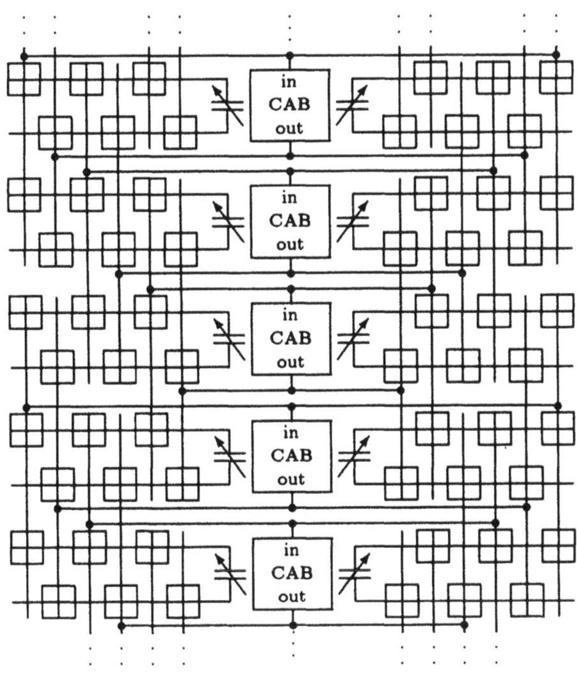

Fig. 10. Proposed architecture with single column.

4.1. *Switched-Capacitor Filters*

The proposed architecture is capable of implementing various SC filter structures. As an illustration, the second-order SC filter shown in Fig. 11 can be embedded in the proposed architecture as shown in Fig. 12. The CABs are configured as an opamp with the feedback switches turned off. Different filter transfer characteristics can be programmed by loading different configuration bits into the PCAs. Since the filter transfer characteristics are dependent on the capacitor ratios and clock frequency, the transfer characteristics are very accurate and do not require tuning of circuit parameters as in the case of other FPAA implementation techniques [4]. If some capacitors in the SC filter are larger than the value provided by a single PCA, the proposed architecture allows for the connection of a few PCAs in parallel to provide the required capacitance value. As illustrated in Fig. 12, capacitors E and F can be realized by PCAs E and E′ and PCAs F and F′, respectively. Following this approach, the resolutions of capacitors E and F are increased by almost 1 bit. In general, if a required capacitor is connected to the input and output terminals of the same CAB, 6 PCAs can be connected in parallel and the maximum increase in resolution for the proposed architecture is slightly less than 2.5 bits. However, this feature will not be effective unless the CAD software including synthesis, routing and placement algorithms is capable of mapping the capacitors into different PCAs. The proposed architecture is capable of implementing multiple-phase SC filters such as N-path SC filters, decimation and interpolation filters since the switches can be controlled by different clock signals. In the macro-block level FPAAs, all the features mentioned above are difficult to accommodate.

An interpolation filter has a transfer function given as

$$H(z) = \frac{1}{r}\frac{1 - z^{-4}}{1 - z^{-1}} \tag{1}$$

A SC realization is shown in Fig. 13. The values of the capacitors are $C_B = C_D = \sqrt{r}C_A = \sqrt{r}C_C$. The embedding of the filter into the proposed architecture is shown in Fig. 14. The FPAA architecture was modelled in SPICE. To simulate the effects of

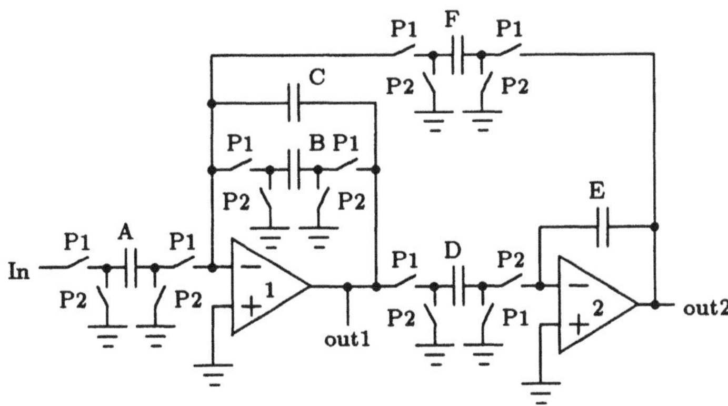

Fig. 11. A second-order SC filter.

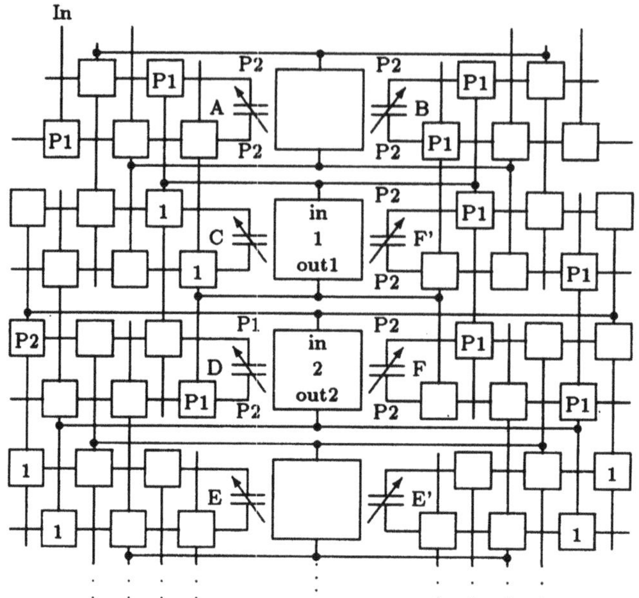

Fig. 12. Embedding of the second-order SC filter.

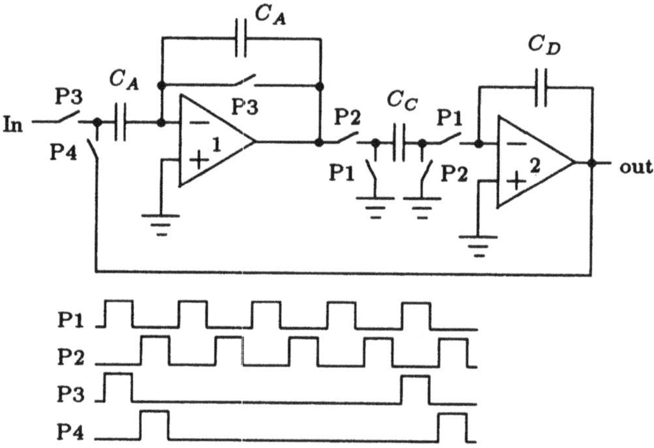

Fig. 13. A SC interpolation filter.

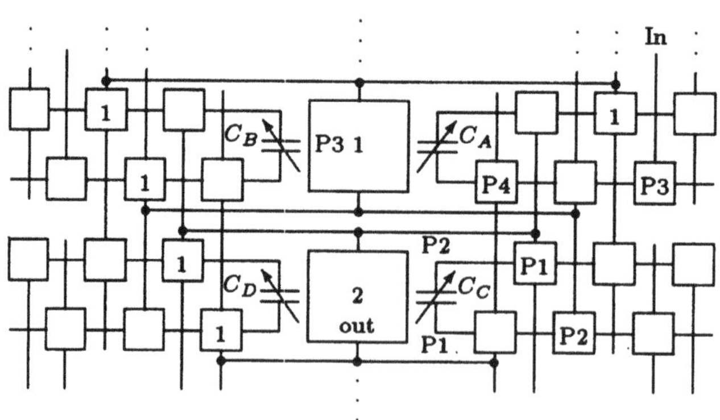

Fig. 14. Embedding of the interpolation filter.

parasitic capacitances, a pair of 0.1 pF capacitors were added to each end of the connection cells. The embedded interpolation filter was simulated based on this model and the simulation results are shown in Fig. 15. In the simulation, r was set to 4 and C_B was set to 1 pF. The unity gain frequency of the opamps was assumed to be 50 MHz. Since the architecture is stray insensitive, the parasitic capacitances have no significant effects on the transfer function given in equation 1. A direct realization of the interpolation filter (not embedded in the FPAA) was also simulated. No significant differences are observed between the embedded circuit and the direct realization circuit.

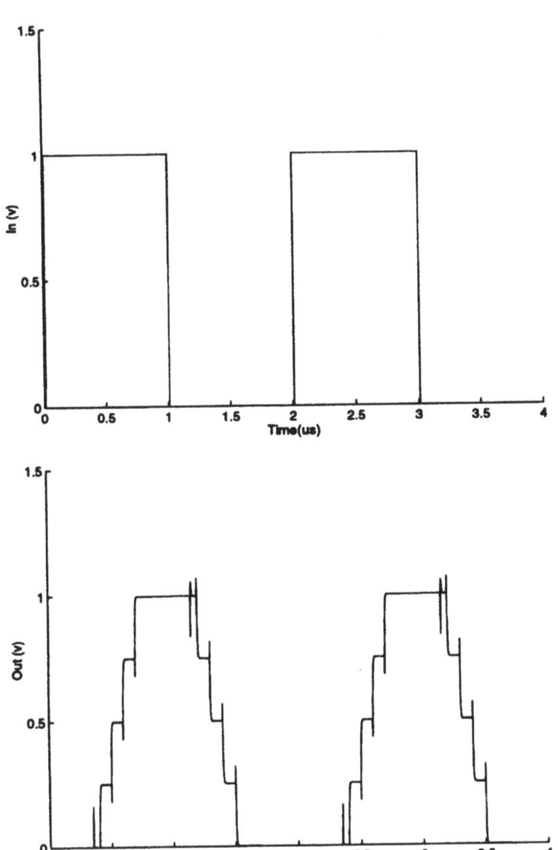

Fig. 15. Simulation results for the embedded interpolation filter.

4.2. *Voltage Controlled Oscillator*

The proposed architecture is capable of implementing different kinds of oscillators including voltage controlled oscillators (VCOs). Fig. 16 and Fig. 17 shows a VCO based on the relaxation oscillator principle [14], and its embedding in the FPAA architecture, respectively. The multiplexer is implemented by programming one of the output connection cells to be controlled by the output of CAB'2 and the clock signal P1. The auxiliary inputs of the output connection cell is connected to the control voltage v_c. The values of the capacitors are set to $C_A = \alpha_0 C_B$, $C_C = \alpha_2 C_B$ and $C_D = \alpha_1 C_B$. Assuming that the absolute values of the maximum and minimum output voltages of opamp 2 (comparator) are equal to V_m, the oscillation frequency f_o for $\alpha_2 << \alpha_1$ is given as

$$f_o = \frac{\alpha_2}{4\alpha_1} f_c + \frac{\alpha_0 v_c}{4\alpha_1 V_m} f_c$$

where f_c is the clock frequency for P1 and P2. The first term in the above equation represents the free-run frequency when v_c is zero. The VCO embedded in the FPAA was simulated and the results are shown in

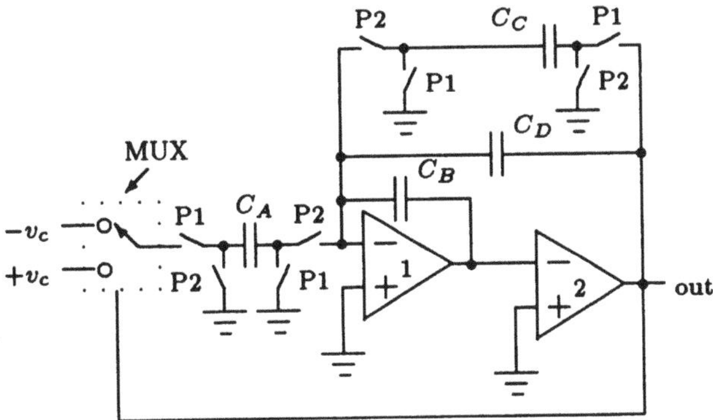

Fig. 16. A VCO based on relaxation oscillator principle.

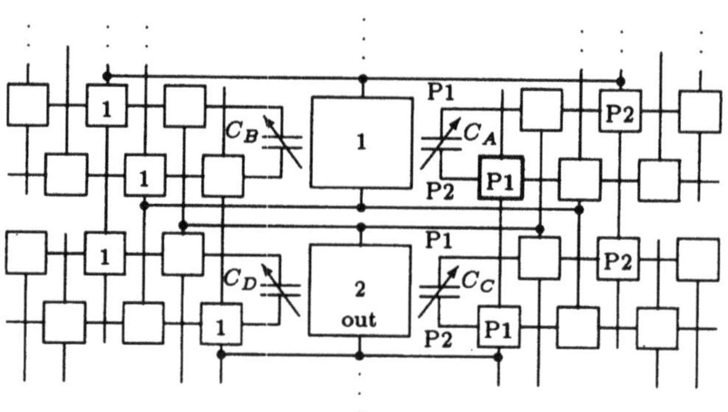

Fig. 17. Embedding of the VCO.

Fig. 18. Results show that f_o changes from about 300 kHz to about 500 kHz for v_c changes from -0.5 V to 0.5 V. The VCO is capable of having a free-run frequency that extends up to a few MHz for f_c in the range of tens of MHz.

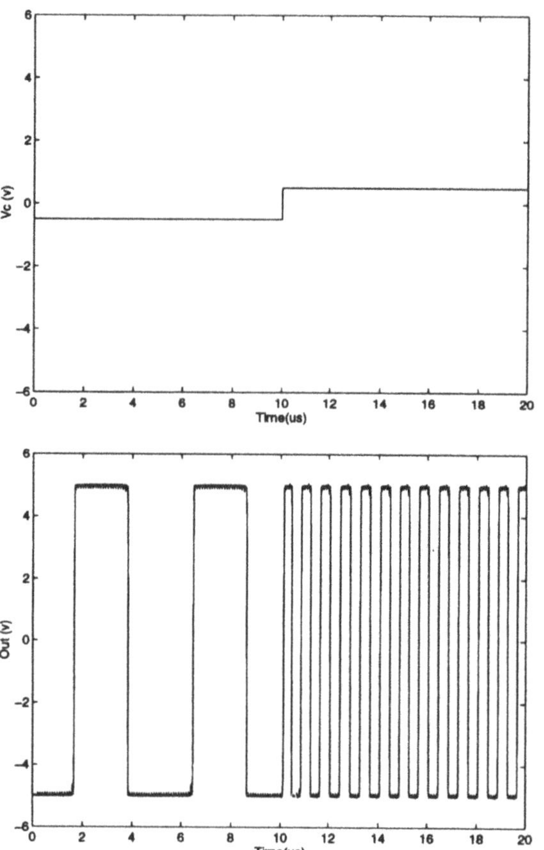

Fig. 18. Simulation results for the embedded VCO.

4.3. Sigma-Delta Modulators

Sigma-delta modulators are usually realized using SC techniques. Fig. 19 shows the SC realization of a second-order modulator with an offset cancellation scheme for the comparator. The circuit can be embedded into the FPAA as shown in Fig. 20. The comparator in Fig. 19 is realized by programming one of the CABs as a comparator with the feedback switches controlled by the clock signal P1. The 1-bit D/A converters are implemented by programming two of the output connection cells to be controlled by the comparator and the clock signal P1 with the auxiliary inputs (referred to Fig. 7) connected to the references $+V_{ref}$ and $-V_{ref}$. Other modulator architectures including high order modulators can be implemented in the FPAA. If a modulator requires a multi-bit D/A converter, this D/A converter can be implemented in the FPAA as discussed in the next section.

4.4. Digital-to-Analog Converters

Various D/A techniques including charge redistribution and algorithmic techniques can be embedded in the proposed architecture if additional logic circuits are added. Fig. 21 shows a charge redistribution design. If the resolution of the D/A converter is less than or equal to the resolution of a PCA, the D/A converter can be implemented using two PCAs and a CAB as indicated in the figure. Since the PCAs in the FPAA can be individually programmed through a data bus, this data bus can be used as the digital inputs (b's) for the D/A converter.

An iterative algorithmic D/A converter can be realized using a sample-and-hold circuit (S/H) and an

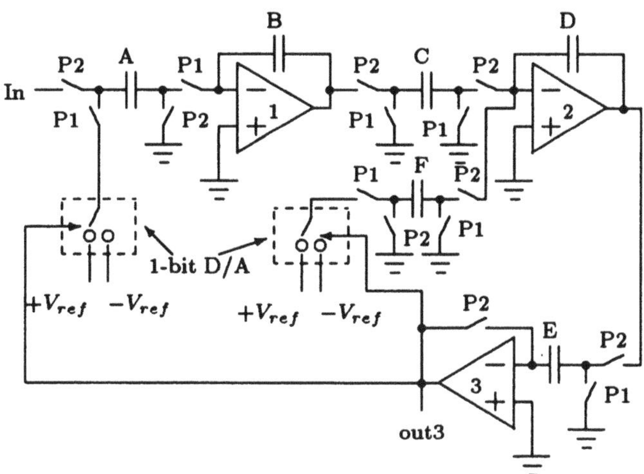

Fig. 19. A second-order sigma-delta modulator.

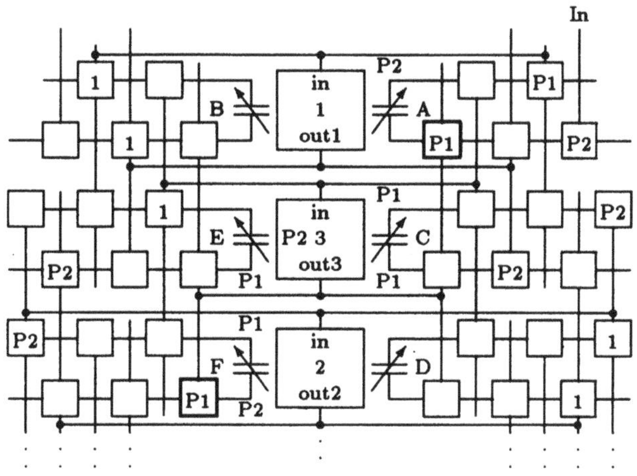

Fig. 20. Embedding of the second-order sigma-delta modulator.

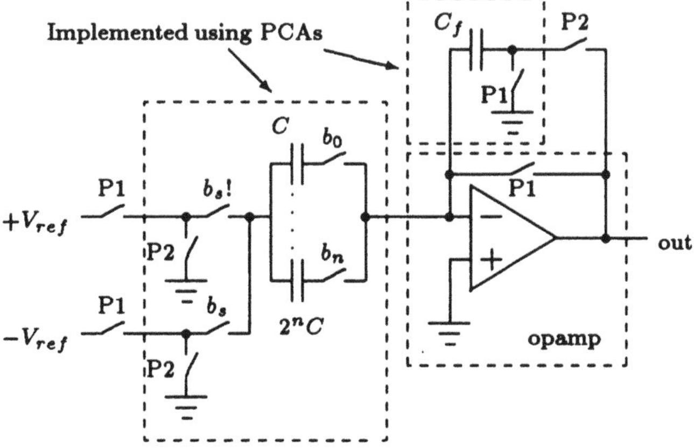

Fig. 21. A charge redistribution D/A converter.

amplifier as shown in Fig. 22 where b_k is the digital input. The conversion is assumed to be started with the LSB (b_0). The embedding of this D/A converter into the FPAA is shown in Fig. 23. The clock signals u and v are assumed to be generated by additional logic circuits which can easily be obtained from a FPGA or from other external circuits. A 4-bit D/A converter embedded in the FPAA was simulated using SPICE. The results are shown in Fig. 24. Only three digital input values (1, 15 and 6) are shown. The output voltage, out, before the reset signal turned high represents the analog output for a given digital input value. In the simulation, the capacitors were assumed to be match. In this case, the D/A converter has high accuracy (better than 10 bits). In practical situations, the D/A converter will be limited to about 10-bit accuracy due to capacitor mismatch.

An iterative D/A converter that starts with the MSB (b_n) can also be embedded in the FPAA. If higher throughput is required, the iterative converter shown in Fig. 22 can be extended to a pipelined D/A converter. A throughput of a few M sample/s to tens of M sample/s can be expected. However, more CABs and PCAs are required in this case.

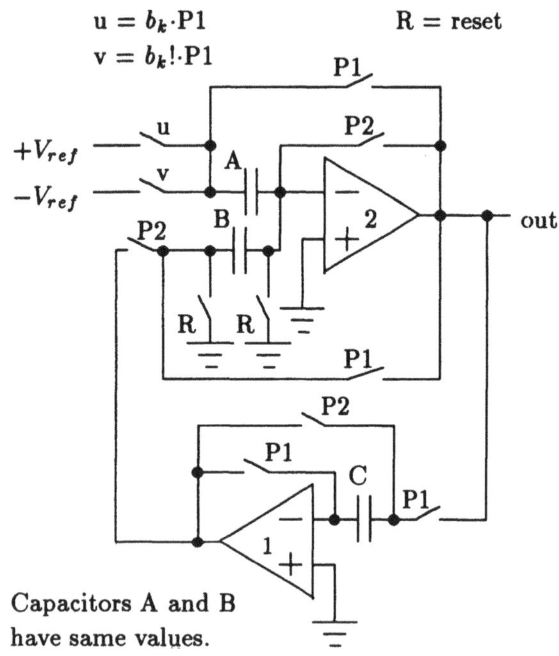

Fig. 22. An iteractive algorithmic D/A converter.

4.5. *Analog-to-Digital Converters*

A/D converters can be implemented in several ways using the proposed FPAA architecture with external logic circuits. An A/D architecture that utilizes a D/A converter is illustrated in Fig. 25. For digital-ramp-type A/D converters, the external logic is a binary counter. For successive-approximation A/D converters, the external logic is a successive-approximation register with additional control logic. The D/A converter in the figure can be implemented using the FPAA as discussed in the previous section. The S/H-and-comparator block can also be embedded in the FPAA using one CAB and one PCA. If a high conversion rate is required, the pipelined A/D converter shown in Fig. 26 can be used. The external logic circuits are digital latches in the simplest case. The SC implementation of a single stage with 1-bit A/D and D/A converters is illustrated in Fig. 27. Each stage consists of a comparator and an amplifier with a

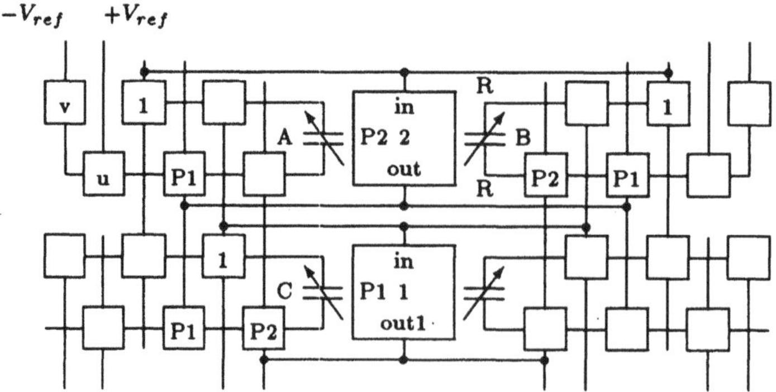

Fig. 23. Embedding of the iterative algorithmic D/A converter.

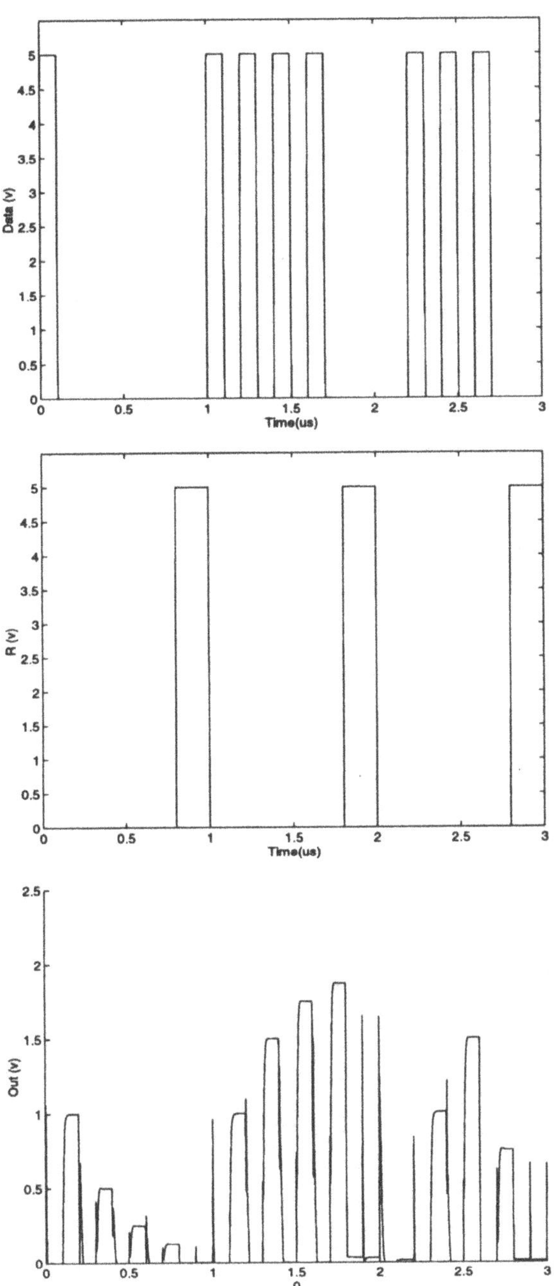

Fig. 24. Simulation results for the embedded D/A converter.

gain of two. They can be embedded in the FPAA as shown in Fig. 28. The 1-bit D/A converter is implemented using the output connection cell with the auxiliary inputs connected to reference sources $+V_{ref}$ and $-V_{ref}$. If high accuracy is required, digital calibration techniques [16] can be used. This does not degrade the speed of the converter nor increase the complexity of the A/D converter realized in the FPAA but does increase the complexity of the external logic circuits. Other A/D architectures such as iterative algorithm A/D converters and integrator-type A/D converters can be implemented in the FPAA. Embedding of companding A/D converters based on μ-law and A-law is achievable using the proposed architecture by combining a comparator, an D/A converter and an A/D converter with external logic circuits.

5. Conclusion

A flexible SC based FPAA architecture was proposed. The architecture utilizes connection switches in the interconnection network as the switches required in SC circuits. Since the number of switches between PCAs and CABs are greatly reduced, the architecture can have a fast settling time and can reach the speed of SC circuits based on custom design. The proposed FPAA provides for the connection of several PCAs in parallel thus reducing the required resolution in the PCAs and the area of the PCAs. Circuit examples were demonstrated in a single column architecture but the architecture is extendible to multi-column architectures. Since useful SC circuits are obtained by local interconnection of components, a local interconnection network is used for interconnecting the PCAs and CABs. The area requirement for such a network was shown to be linearly proportional to the number of CABs or PCAs. The overall architecture is area efficient due to the savings of area from the PCAs and the linear growth properties on the interconnection network. The local connections in the interconnection network also reduce the parasitic capacitances and correspondingly increase the speed of operations and reduce the cross-talk between connections. Various useful circuits including SC filters, VCOs, sigma-delta modulators and different types of A/D and D/A converters can be implemented as demonstrated in this paper. Other linear circuits such as programmable gain amplifiers can also be implemented in the proposed architecture. Since the FPAA can realize D/A and A/D converters, it can be directly integrated with FPGAs for realizing field-programmable mixed analog and digital arrays which can be used for prototyping mixed signal circuits.

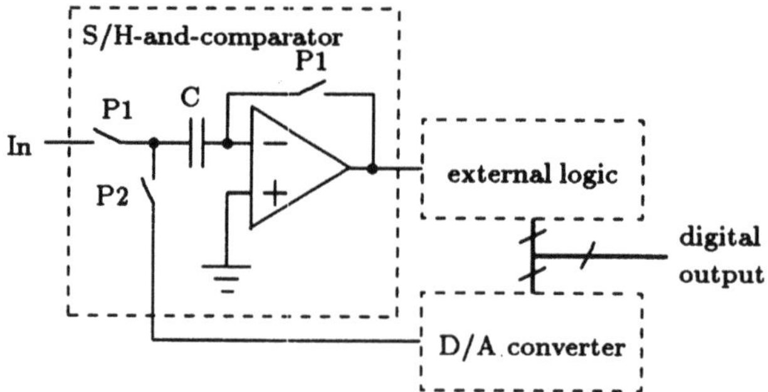

Fig. 25. A/D converter that uses D/A converter.

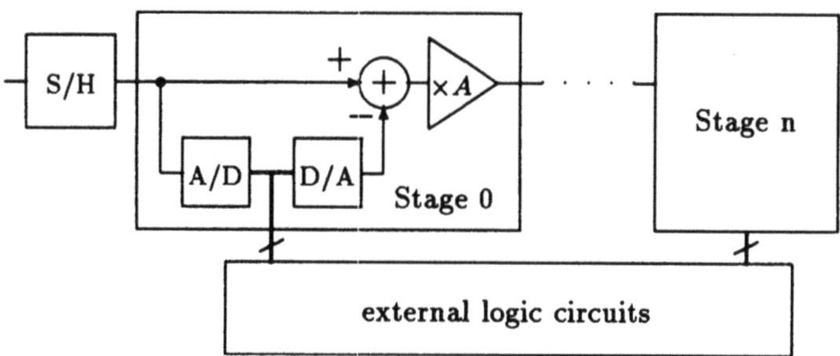

Fig. 26. Pipelined A/D converter architecture.

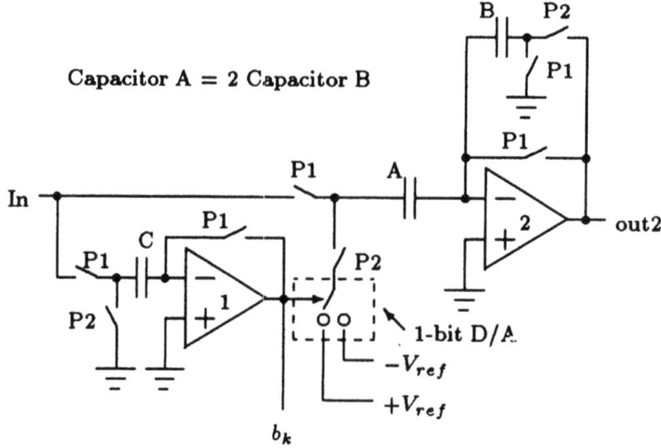

Fig. 27. SC realization of each pipelined stage using 1-bit A/D and D/A converters.

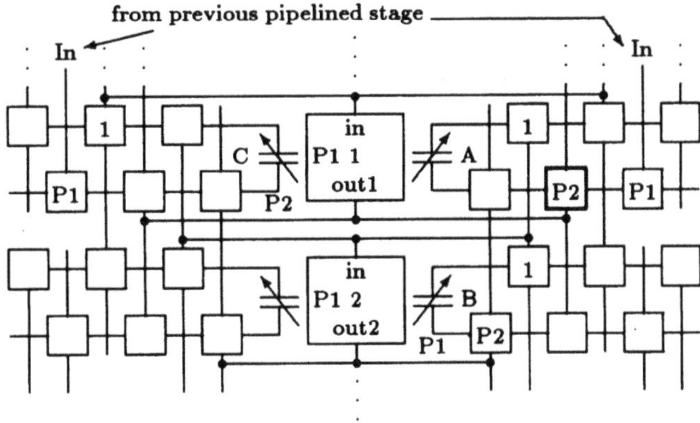

Fig. 28. Embedding of one pipelined stage.

Acknowledgment

The authors would like to thank Prof. R. Geiger and Prof. S. Sapatnekar for their valuable comments and suggestions.

Notes

1. To guarantee satisfactory settling behavior, the unity gain frequency of the opamps has to be at least 5 times greater than the clock rate [14].
2. The symbol "!" represents the complement of a logic signal in this paper.
3. The parasitic capacitance on an interconnection wire is smaller than the capacitance normally programmed to a PCA.

References

1. E. K. F. Lee and P. G. Gulak, "A CMOS Field Programmable Analog Array." *IEEE ISSCC Digest of Technical Papers 1991*, pages 186–187, 1991.
2. E. K. F. Lee and P. G. Gulak, "A CMOS Field Programmable Analog Array." *IEEE Journal of Solid State Circuits* 26(12), pp. 1860–1867, Dec. 1991.
3. E. K. F. Lee and P. G. Gulak, "Field Programmable Analogue Array Based On MOSFET Transconductors." *Electronics Letters* 28(1), pp. 28–29, Jan. 1992.
4. E. K. F. Lee and P. G. Gulak, "A Transconductor Based Field Programmable Analog Array." *IEEE ISSCC Digest of Technical Papers 1995*, pp. 198–199, 1995.
5. E. Pierzchała, M. A. Perkowski, and S. Grygiel, "A Field-Programmable Analog Array for Continuous, Fuzzy, and Multi-Valued Logic Applications." *Proceedings of the Twenty-Fourth International Symposium on Multiple Valued Logic*, pp. 148–155, May 1994.
6. E. Pierzchała, M. A. Perkowski, P. Van Halen, and R. Schaumann, "Current-Mode Amplifier/Integrator for a Field-Programmable Analog Array." *IEEE ISSCC Digest of Technical Papers 1995*, pp. 196–197, 1995.
7. M. A. Sivilotti, "A Dynamically Configurable Architecture for Prototyping Analog Circuits." In *Advanced Research in VLSI: Proceedings of the Decennial Caltech Conference on VLSI, 1988*. MA: MIT Press, 1988.
8. H. Kutuk and S. Kang, "A Field-Programmable Analog Array (FPAA) Using Switched-Capacitor Techniques." *IEEE ISCAS*, 3, pp. 41–44, 1996.
9. F. Goodenough, "Analog Counterparts of FPGAs Ease System Design." *Electronic Design*, Oct. 1994.
10. H. Klein, "The EPAC Architecture: An Expert Cell Approach to Field Programmable Analog Devices." *ACM/SIGDA FPGA'95*, Monterey, CA, pp. 94–98, Feb. 12–14, 1995.
11. A. Bratt and I. Macbeth, "Design and Implementation of a FPAA." *ACM/SIGDA FPGA'95*, Monterey, CA, pp. 88–93, Feb. 12–14, 1995.
12. C. Zhang, A. Bratt, and I. Macbeth, "A New Field Programmable Mixed Signal Array and Its Application." *4th Canadian Workshop on Field-Programmable Devices*, Toronto, Canada, May 13–14, 1996.
13. P. Chow, P. Chow, and P. G. Gulak, "A Field Programmable Mixed Analog-Digital Array." *ACM/SIGDA FPGA'95*, Monterey, CA, pp. 104–109, Feb. 12–14, 1995.
14. R. Gregorian and G. C. Temes, *Analog MOS Integrated Circuits for Signal Processing*. John Wiley & Sons, 1986.
15. P. V. Ananda Mohan, V. Ramachandran, and M. N. S. Swamy, *Switched Capacitor Filters: Theory, Analysis and Design*. Prentice Hall, 1995.
16. E. G. Soenen and R. L. Geiger, "An Architecture and an Algorithm for Fully Digital Correction of Monolithic Pipelined ADC's" *IEEE Trans. on Circuits and Systems—II: Analog and Digital Signal Processing* 42(3), 143–153, Mar. 1995.

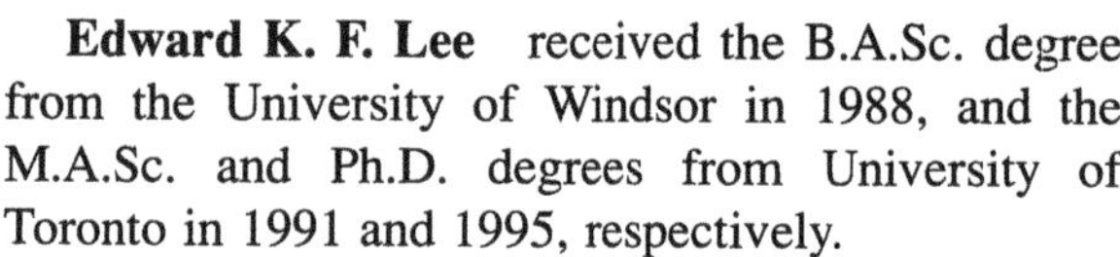

Edward K. F. Lee received the B.A.Sc. degree from the University of Windsor in 1988, and the M.A.Sc. and Ph.D. degrees from University of Toronto in 1991 and 1995, respectively.

Since 1995, he has been an Assistant Professor of electrical and computer engineering at Iowa State University. His current research interests are in the area of analog and digital VLSI design and field-programmable mixed analog and digital arrays.

Wai Leung Hui received the B.S. degree in Electrical Engineering in 1996 at Iowa State University, Ames, Iowa, U.S.A. Currently, he is working on integrated GMR isolation technique for his Master degree research project.

Analog Integrated Circuits and Signal Processing, 17, 51–65 (1998)

A Switched Capacitor Approach to Field-Programmable Analog Array (FPAA) Design

HAYDAR KUTUK AND SUNG-MO (STEVE) KANG

Department of Electrical and Computer Engineering and Coordinated Science Laboratory, University of Illinois at Urbana-Champaign-USA
hkutuk@uivlsi.csl.uiuc.edu

Received July 1, 1996; Accepted May 23, 1997

Abstract. We present a voltage mode switched-capacitor Field Programmable Analog Array (FPAA) to be used to implement various analog circuits. The FPAA consists of uniform configurable analog blocks (CABs) allowing implementation of different functions. Each CAB consists of two back-to-back connected inverting and non-inverting strays-insensitive switched-capacitor integrators. The interconnection between CABs is implemented by switched and unswitched capacitor networks. The internal structure of CABs and the interconnection between different CABs are configured by user-programmable digital control signals. The functionality of the FPAA is demonstrated through embedding of different types of filters, programmable amplifiers, biquads, modulators and signal generators along with simulation results.

Key Words: switched-capacitor, configurable analog block, field-programmable analog array

1. Introduction

In spite of rapid advances in digital programmable gate arrays, namely field programmable gate arrays (FPGAs), there has been little improvement on their analog counterparts. There exists a significant need for an architecture of programmable analog circuit array which provides flexibility for fast prototyping and short turnaround times.

Previously, three attempts have been made to solve this problem. In the first approach [1] the programmability feature of configurable analog blocks (CABs) was achieved by user programmable switches placed into the signal paths. In order to reduce the nonlinear and distortion effects of the switches, MOS transistors were operated in the sub-threshold regions. In the second approach [2] differential MOS transconductances were used as programmable resistors and the programmability was achieved by varying the values of the transconductances. In the third approach [3] continuous-time current-mode circuits were used to achieve higher performance. The programmability was achieved by changing the bias currents in the translinear bipolar topology.

In this paper we propose switched-capacitor (SC) techniques and address the impact of the techniques on the programmability and reconfigurability[1] of CABs and the interconnection network between CABs [4] and [5]. We chose SC circuits in our FPAA for the following reasons. First, SC circuits are readily suited for programming and reconfiguration because of the existing switches which can be used for programming and reconfiguration. Switches used in SC networks must be distinguished from the switches used for programming. Programming switches either connect or disconnect additional elements to the element to be programmed. Switches in SC circuits form an integral part of the circuit and they act as sampling elements activated by non-overlapping clock phases. They transfer charges on a particular set of capacitors to other capacitors. Even though these switches are in series connection with capacitors, they only affect the settling time [6], needed for transferred charges to reach their final values. Consequently, the programming and reconfiguration of capacitors switched by non-overlapping clock phases, can be achieved without using any additional switches. In order to program the values of unswitched capacitors, we use programming switches such as the programmable capacitor array (PCA) [7]. Second, by altering the clock frequency, effective capacitor values can be changed without changing their physical dimensions. Third, SC circuits can be designed very accurately by adjusting the ratio of capacitors which can be achieved in a CMOS process. The voltage-mode operation and the use of op-amps provide nearly ideal summing of different signals. More importantly, no complicated on-chip tuning circuit is necessary as is the case in current-mode and other continuous time applications. However, the sampled-data nature of the SC circuits requires pre- and post-processing filters for anti-aliasing and reconstruction (smoothing).

We will describe the internal structure of the CAB and the interconnection network among different CABs in Section 2. The control circuit and user interface will be explained in Section 3, followed by Section 4 which will show how different types of filters can be embedded into the system. In Section 5 non-filtering applications are demonstrated. Simulation results for several design examples will be presented.

2. Configurable Analog Block and Interconnection Network

The schematic of the configurable analog block (CAB) is given in Fig. 1. The CAB can be viewed as a functional block with two input ports and two output ports with the inclusion of control signals and non-overlapping clock phases. The CAB consists of back-to-back connected strays-insensitive inverting and non-inverting integrators [8]. One of the two integrators is lossy and the other one is lossless. Three types of digitally controlled configuration schemes are provided in the CAB.

1. Phase programming is used to provide inverting or non-inverting integration depending on the application. It is achieved by interchanging the phases of non-overlapping clock signals at the switches.
2. Connection and disconnection scheme is used either to increase the circuit functionality by adding new passive or active elements or to disconnect unnecessary elements from the circuit.
3. Capacitor value programming provides four different capacitor values. Capacitors are connected in parallel to give the desired value.

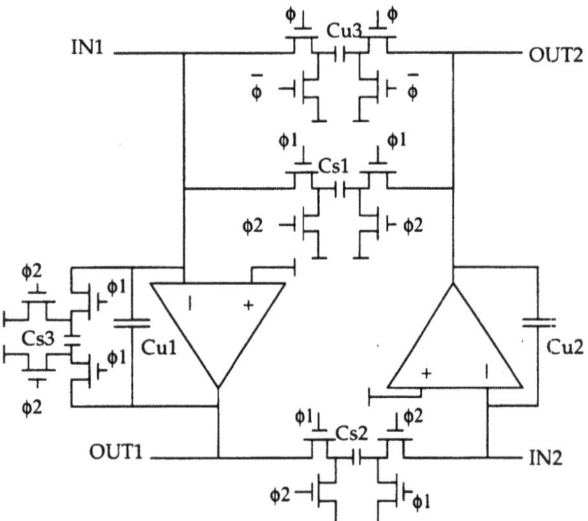

Fig. 1. The configurable analog block schematic.

Depending on the application, the integrator loop in the CAB can be broken into separate integrators by using the connection/disconnection scheme. The op-amps are connected via phase inverting and non-inverting SC networks representing positive and negative resistors, respectively. Simply enabling the clock phases to the MOS transistors operating as switches will activate the connection between the two integrators. Disabling of phases will disconnect op-amps. In addition to the integrators, the positive resistor simulating switched capacitor network is bridged by an unswitched capacitor, C_{u3}, providing more flexibility for further applications such as biquad synthesis. The complexity of the CAB can be increased at the expense of increasing the complexity of the digital control circuit. The op-amps are designed as class-AB [9] having a DC gain around 90 dB and 60° phase margin so that they would be suitable for most applications.

The connection among CABs is provided by the interconnection network which is shown in Fig. 2. The CABs can be connected to each other in two ways: by switched capacitors if the application requires a summing operation at the particular node, or by an unswitched capacitor[2] if the particular application requires the realization of a finite transmission zero. Thus, the connection network must provide a means for both types of applications. This doubles the number of connections. In order to reduce the number of connection networks without affecting the programming flexibility of the interconnection network too much, we limit the connections to 4 unswitched capacitors instead of 8 and to 4 switched capacitors (2 for simulating positive resistors and 2 for simulating negative resistors) instead of 8. The control

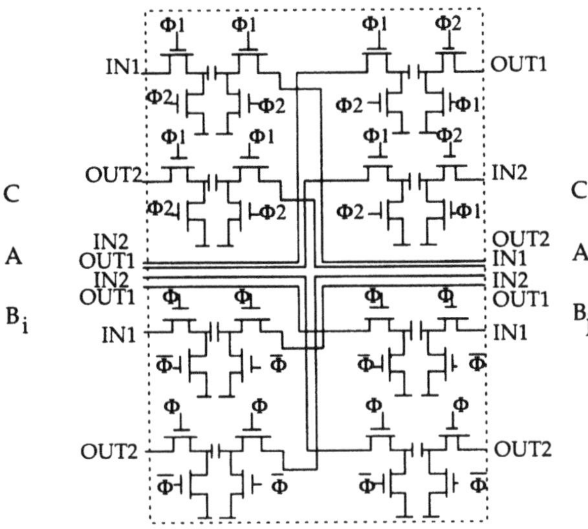

Fig. 2. The programmable interconnection network between two CABs.

of the interconnection network is similar to the internal control of the CAB. When the control circuit enables the clock phases to the desired interconnection switched or unswitched capacitor, a connection is established. Disabling the clock phases disconnects the desired interconnection. Since the switches are already present for the switched capacitor interconnection, no additional switches are introduced into the interconnection networks unlike in previous approaches in the literature.

The overall FPAA structure including the anti-aliasing and reconstruction filters and the digital control circuit is shown in Fig. 3. The structure can be repeated for more complicated circuits.

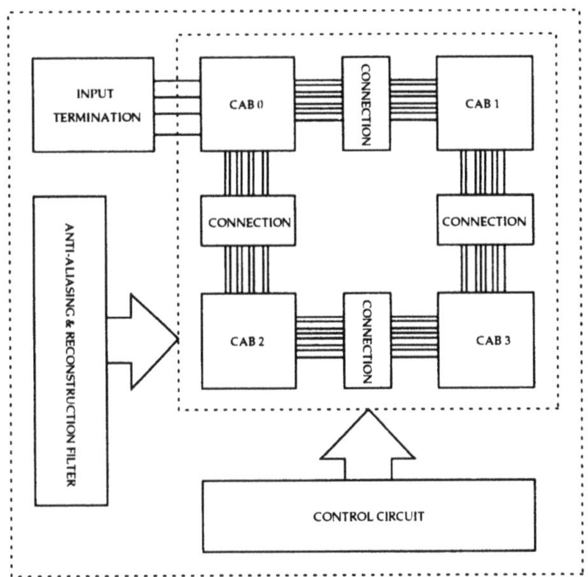

Fig. 3. The overall FPAA structure: all configurable four CABs and interconnections in between.

3. FPAA Digital Control Circuit

The digital control circuit is designed using the algorithmic state (ASM) machine approach. The control circuit consists of a data processor subsystem manipulating the data and a control logic generating the control signals sequencing the desired operations in the data processor. The control sequence and data processing tasks are described as a hardware algorithm in Fig. 4. The user-programmable configuration bits, given in Fig. 5, are stored as control words and downloaded either from a PC or a preprogrammed EPROM. The data processor consists of three registers and an address counter. The information encoded in the user-defined control words is decoded and the control signals are sent to the appropriate parts in the architecture. The decoding logic circuit is given in Fig. 6. Each step of the algorithm is coded as a state as shown in the ASM chart. The sequencing of these steps is controlled by the control logic, which is designed in three levels: MUX(4-inputs, 2-control inputs), D flip-flop and decoder(1:4) providing a very regular architecture. The overall FPAA digital-control circuit including the control-logic part and the data processor-logic are shown in Fig. 7.

4. Embedding Filters into the FPAA

For filter design applications, doubly terminated LC ladder filters are used as prototype circuits. The continuous-time filters are approximated by an s-domain transfer function, which is then converted into the z-domain transfer function by applying the bilinear transformation. The realization of this z-domain transfer function yields the exact design [10]. The variables used in exact design are as follows,

$$\lambda = \frac{s_{ct}T}{2} = \frac{z^{1/2} - z^{-1/2}}{z^{1/2} + z^{-1/2}} = \tanh(sT/2) \qquad (1)$$

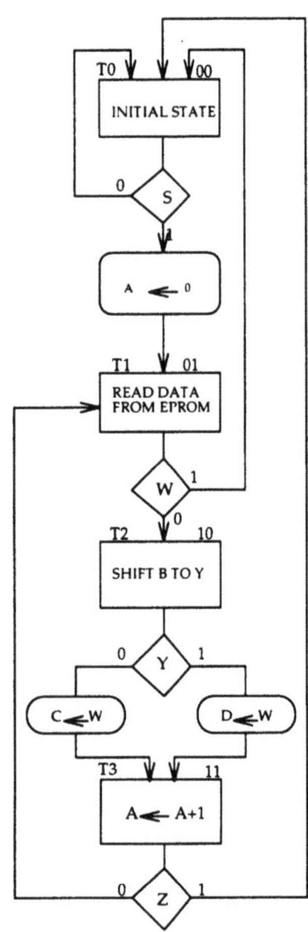

Fig. 4. Algorithmic state machine flow chart of the control logic.

$$\gamma = \frac{1}{2}\left(z^{1/2} - z^{-1/2}\right) = \sinh(sT/2) \tag{2}$$

$$\mu = \frac{1}{2}\left(z^{1/2} + z^{-1/2}\right) = \cosh(sT/2) \tag{3}$$

The bilinear transformation is defined by λ. The integrators used in the FPAA realize the following transfer function,

$$\pm c\frac{1}{\gamma} = \pm c\frac{1}{\sinh(sT/2)} \tag{4}$$

In order to realize the λ-domain circuit, all impedances must be scaled by dividing them by $\mu = \cosh(sT/2)$ and all admittances must be scaled by multiplying them by μ. After scaling the circuit is in γ-plane. In the filter design we used three different types of subblocks. For the first subblock, as shown in Fig. 8, the transfer function in γ-plane is

$$v_0 = -\frac{1}{\mu g_k + \gamma C_1}(ai_1 + bi_2) \tag{5}$$

This equation can be implemented by an active-RC circuit and finally it is converted into an active-SC realization. The transfer function of its active-SC realization is

$$v_0 = -\frac{C_1 v_1 + C_2 v_2}{C_0\mu + (2C_{int} + C_0)\gamma} z^{1/2} \tag{6}$$

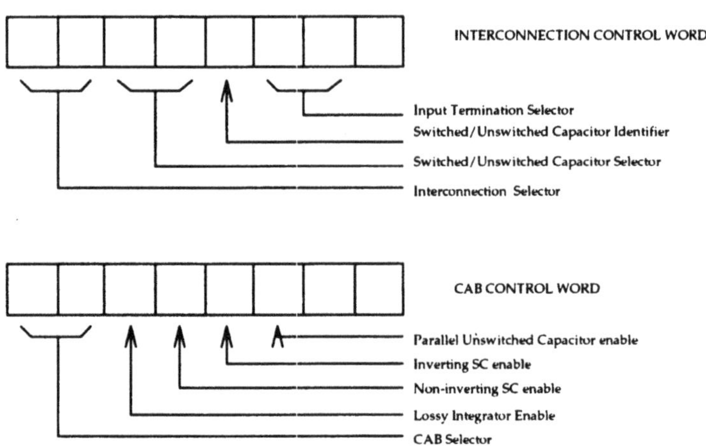

Fig. 5. User-defined control words for CAB and interconnection reconfiguration.

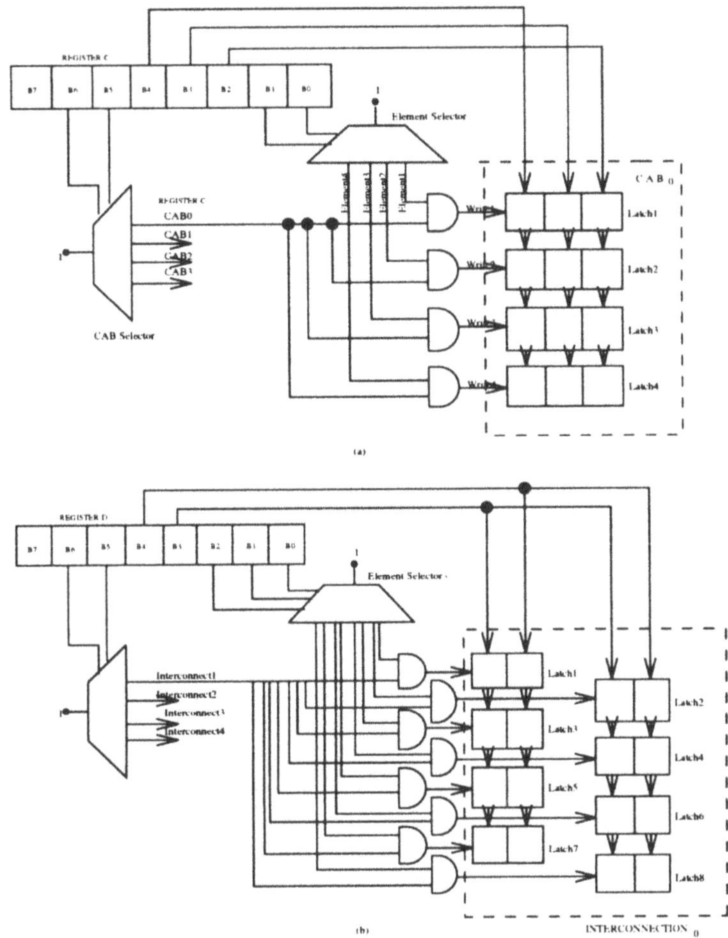

Fig. 6. Decoding of user-defined information on registers.

The other commonly used subblock is shown in Fig. 9 and its transfer function is

$$v_0 = -\frac{1}{\mu g_k \gamma\left(C_1 + \frac{1}{L_1}\right) + \frac{1}{\gamma L_1}}(ai_1 + bi_2) \tag{7}$$

This circuit is also converted into an active-SC circuit with the following transfer function.

$$v_0 \doteq -\frac{C_1 v_1 + C_2 v_2}{C_0\mu + (2C_{int1} + C_0)\gamma + \frac{C_3 C_4}{2C_{int2}\gamma}} z^{1/2} \tag{8}$$

The series branch circuit is shown in Fig. 10 and its λ-plane transfer function is given as

$$i_0 = \left[\gamma\left(C_1 + \frac{1}{L_2}\right) + \frac{1}{\gamma L_2}\right](av_1 + bv_2) \tag{9}$$

By converting this subcircuit into an active-SC equivalent the following transfer function is formulated.

$$v_0 \frac{C_1 v_1 + C_2 v_2}{2C_{int1}\gamma} z^{-1/2} \tag{10}$$

4.1. *Low-pass Filter Implementation*

As a simple example, let us consider a third order low-pass filter. From Fig. 11, it can be seen that this prototype circuit can be built using previously described subblocks. Since the proposed CABs are

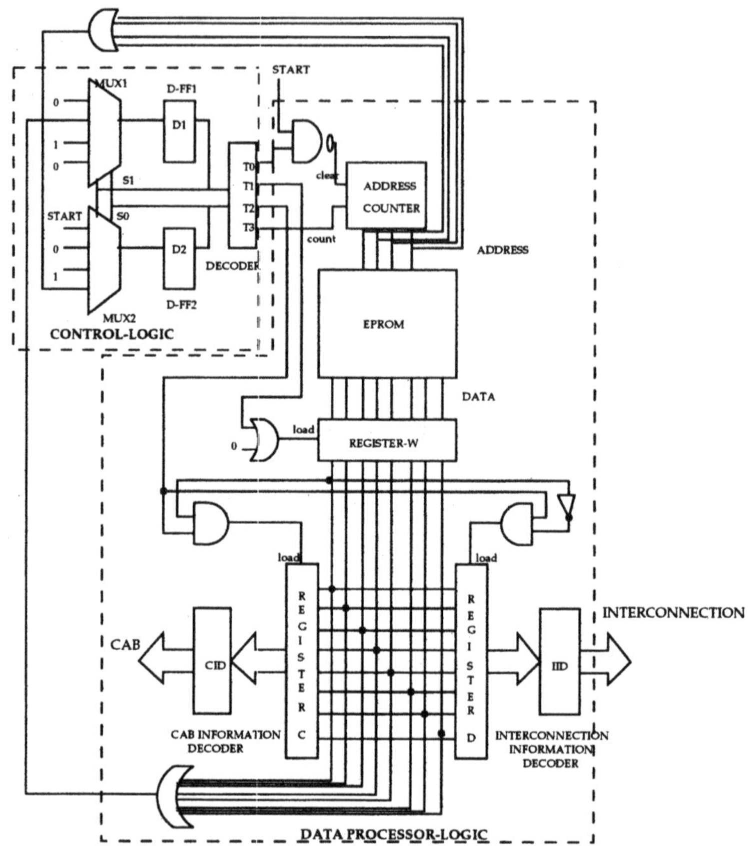

Fig. 7. The overall digital control circuit for the FPAA.

suitable to embed these subblocks, the filter can be designed using two CABs. The γ-plane equations of the LC prototype circuit are given as

$$-v_1 = -\frac{\mu g_s v_s + (-i_2) + \gamma C_{a2} v_2}{\mu g_s + \gamma(C_3 + C_{a2})} \tag{11}$$

$$-i_2 = \frac{(-v_1) + v_2}{\gamma L_2} \tag{12}$$

$$v_2 = -\frac{(-i_2) + C_{a2}(-v_1)}{\mu g_l + \gamma(C_3 + C_{a2})} \tag{13}$$

The active-RC realization can be realized with the following equations

$$-V_1 = -\frac{G_1 R_a V_{in} + G_3 R_a(-V_2) + sC_{n4} R_a V_3}{G_{d1} R_a + sC_{n1} R_a} \tag{14}$$

$$-V_2 = \frac{G_2 R_b(-V_1) + G_4 R_b V_3}{sC_{n2} R_b} \tag{15}$$

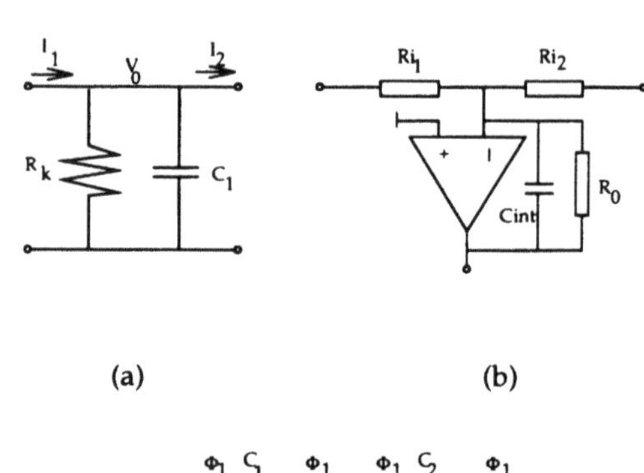

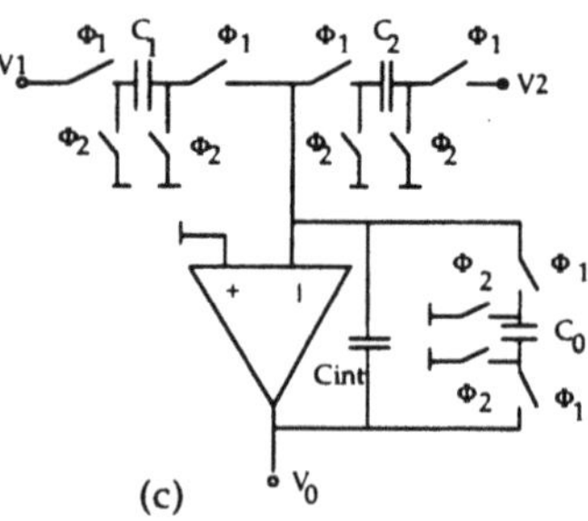

Fig. 8. (a) Basic shunt-branch element I, (b) its active-RC realization, and (c) active-SC realization.

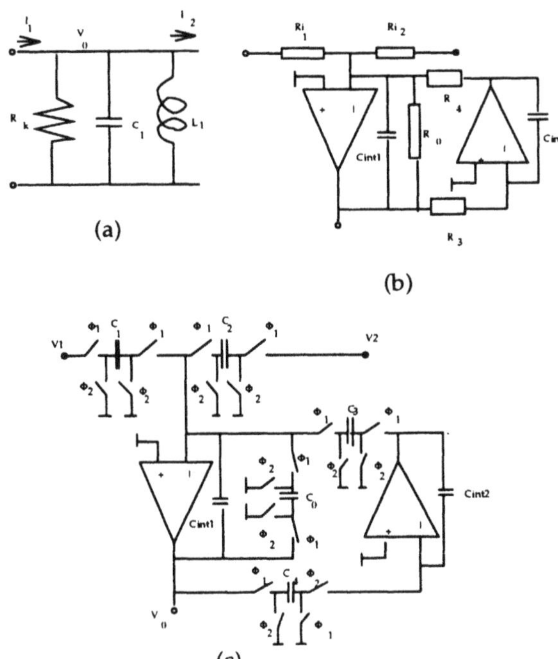

Fig. 9. (a) Basic shunt branch element II, (b) its active-RC realization, and (c) active-SC realization.

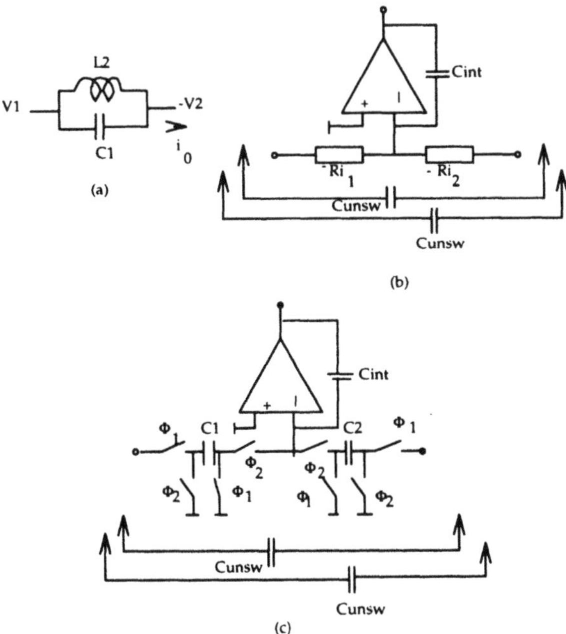

Fig. 10. (a) Basic series branch element, (b) its active-RC realization, and (c) active-SC realization.

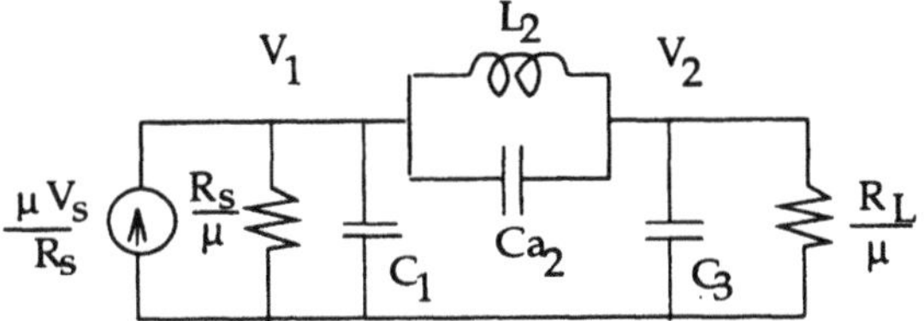

Fig. 11. The γ-plane equivalent prototype circuit for the low-pass filter.

$$V_3 = \frac{G_5R_c(-V_2) + sC_{n5}R_c(-V_1)}{G_{d2}R_c + sC_{n3}R_c} \tag{16}$$

By coefficient comparison of the passive-LC and active-RC circuit, the component values are calculated. Finally, the resistors in the active-RC circuit are converted into switched capacitors. Two CABs are necessary to embed this filter into the FPAA. Since the order is odd, i.e., 3, one of the CABs must be divided into a single lossy integrator using the connection/disconnection scheme. The interconnection between the CABs is formed by two unswitched and two switched capacitors. The embedded filter is shown in Fig. 12, and its simulated frequency response is shown in Fig. 13 along with clock frequency programming. The clock frequency ranges from 2 kHz to 256 kHz.

4.2. *Band-pass Filter Implementation*

The design of a band-pass filter also starts with a passive-LC prototype. By applying frequency transformation techniques the band-pass equivalent is calculated. The low-pass to band-pass transformation is given in Fig. 14. If the prototype in (b) is implemented with the subblocks described earlier, an additional opamp is needed. Therefore, the conversion from (b) to (c) is applied. Once the band-pass filter in Fig. 14(c) is formed, its γ-plane transformation is derived as the following

$$v_1 = -\frac{\mu g_s v_s + \frac{L_3}{L_2} i_3 + \left(1 + \frac{L_1}{L_2}\right)(-i_1) + \gamma C_{a2} v_2}{\mu g_s + \gamma(C_{a1} + C_{a2})} \tag{17}$$

$$-i_1 = -\frac{v_1}{\gamma L_1} \tag{18}$$

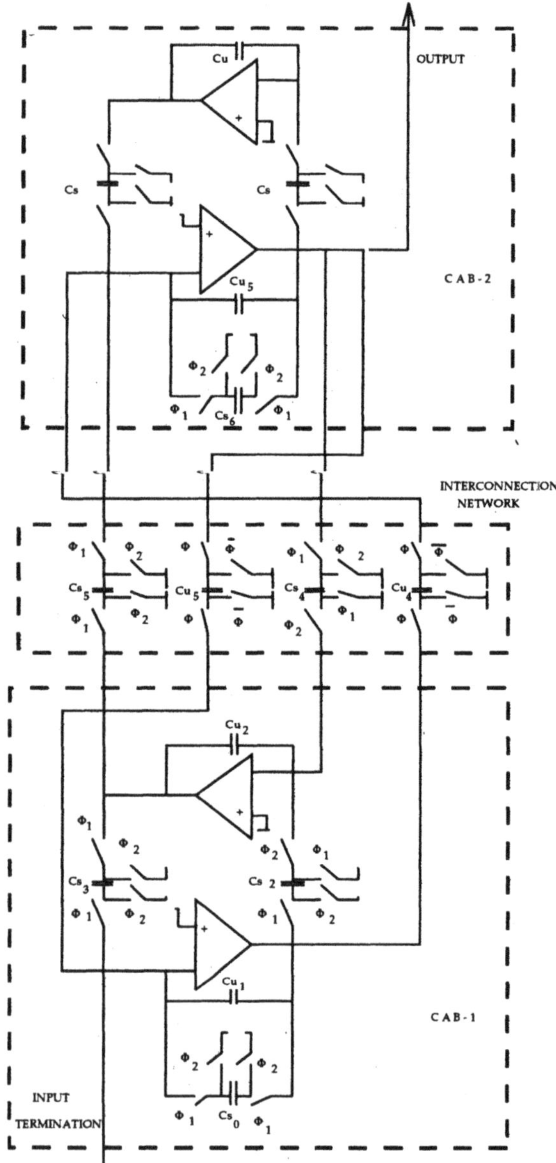

Fig. 12. Embedding of a third-order low-pass elliptic switched-capacitor filter.

$$v_2 = -\frac{\frac{L_1}{L_2}(-i_1) + \left(1 + \frac{L_3}{L_2}\right)i_3 + \gamma C_{a2}(-v_1)}{\mu g_L + \gamma(C_{a2} + C_{a3})} \tag{19}$$

$$i_3 = \frac{v_2}{\gamma L_3} \tag{20}$$

The equations corresponding to the active-RC circuit are calculated similarly to the low-pass filter case and they are given as

$$-V_1 = \frac{G_1 R_a V_{in} + G_3 R_a(-V_4) + G_5(-V2) + sC_5 R_a V_3}{G_0 R_a + sC_1 R_a} \tag{21}$$

$$-V_2 = \frac{G_2 R_b}{sC_2 R_b}(-V_1) \tag{22}$$

$$V_3 = -\frac{G_7 R_c(-V2) + G_9(-V_4) + sC_6(-V_1)}{G_{10} R_c + sC_3 R_c} \tag{23}$$

$$-V_4 = \frac{G_4 R_d}{sC_4 R_d} V_3 \tag{24}$$

Again by coefficient comparison the component values in the active-RC are determined. The active-RC circuit is converted into the active-SC circuit. Two CABs are needed to embed the fourth-order band-pass filter into the FPAA. The even order of the filter allows the use of each CAB without dividing them into single amplifiers. Assuming a doubly terminated network, two lossy integrators are necessary. These integrators are provided in the CABs. The band-pass filter type explained so far can be directly embedded without any configuration on the CABs. The fourth order requires one interconnection block between the two CABs. The interconnection block must consist of two switched capacitors, C_{s3} and C_{s7}, and two unswitched capacitors, C_{u5} and C_{u6}. The switched capacitors, C_{s3} and C_{s7}, simulate a charge non-inverting capacitor or positive resistor. The configuration of CABs and the interconnection box are shown in Fig. 15. The same approach enables higher-order filter implementations by making proper connections via the user-controllable interconnection block. The frequency response of the fourth-order band-pass filter is shown in Fig. 16 along with clock frequency programming. By programming the clock frequency from 2 kHz up to 256 kHz, the center frequency of the filter can also be varied from 350 Hz up to 45 kHz without changing any physical capacitor sizes.

Higher-order filters are designed in a similar way. An eighth-order band-pass filter is designed and its frequency response is shown in Fig. 17. Four CABs are needed to embed this filter into the FPAA. The frequency response for a fifth-order low-pass filter is shown in Fig. 18. Three CABs are needed to implement this filter using the FPAA architecture. The FPAA with its connection/disconnection scheme

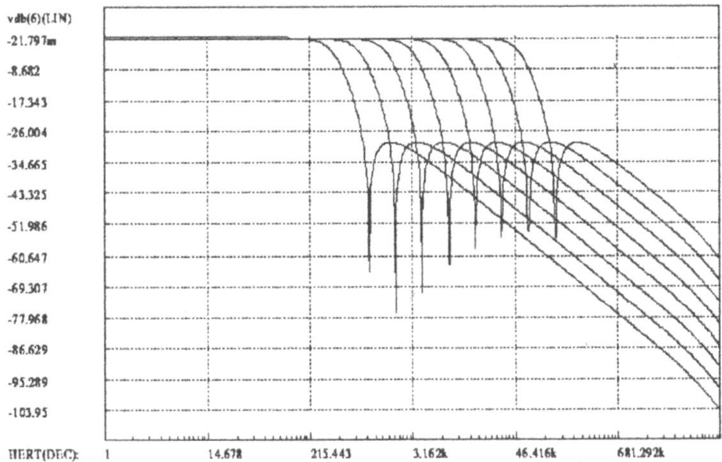

Fig. 13. Frequency response of the third-order low-pass filter with the clock frequency programmed from 2 kHz to 256 kHz.

enables using a higher-order filter as a lower-order filter without going through a new design cycle. As an example, the same fifth-order low-pass filter is configured as a third-order filter by disconnecting one of the CABs. The frequency response for that filter is given in Fig. 19.

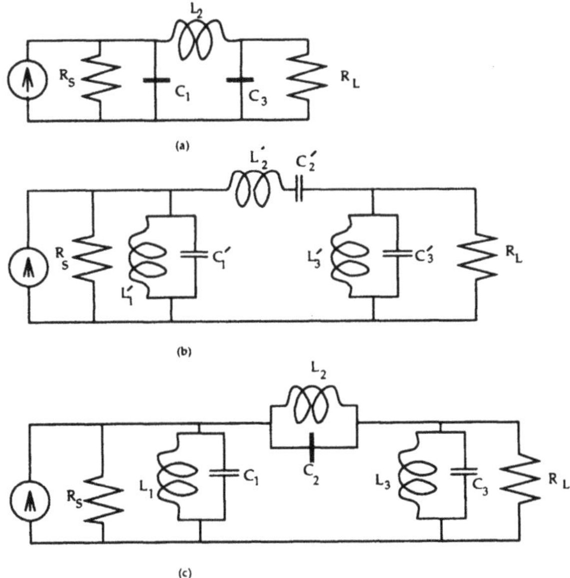

Fig. 14. (a) Low-pass prototype circuit, (b) the low-pass to band-pass transformed circuit, (c) the band-pass circuit for minimum opamp realization.

4.3. Biquad Implementation

A different approach to filter design is cascading biquads. Although this technique is simple to implement, it suffers from high sensitivity to pole locations of each individual biquad. The second-order nature of the CABs allows the implementation of various biquad circuits. Using the regular architecture of the FPAA, the same biquad circuit can be easily cascaded for higher-order circuit implementations. A high-Q biquad [6] has a transfer function of

$$H(z) = -\frac{C_6z^2 + (C_1C_3 + C_5C_3 - 2C_6)z + (C_6 - C_5C_3)}{z^2 + (C_2C_3 + C_3C_4 - 2)z + (1 - C_3C_4)} \tag{25}$$

By comparing this expression with the general biquad expression the capacitor values are calculated. If these values are properly chosen, the biquad can be configured as low-, band- and high-pass. A single high-Q biquad configured as low-pass is embedded into one CAB and its frequency response is shown in Fig. 20.

5. Embedding Non-filtering Applications into the FPAA

In this section we will show that the FPAA can implement analog signal amplifiers, modulators and signal generators.

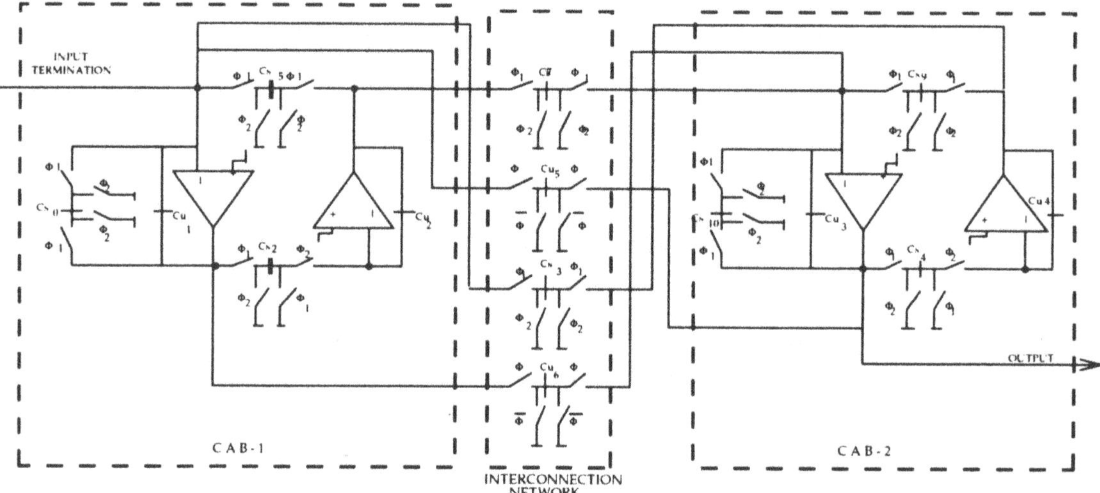

Fig. 15. Embedding of a fourth-order elliptic band-pass switched-capacitor filter.

5.1. *Programmable Amplifier Implementation*

Analog signal amplification is achieved by using switched capacitors in the feedback loop and in the input branch of an opamp. The gain is simply determined by the ratio of the input capacitor and feedback capacitor. Since the switched capacitors will have the same switching phases, in one phase the input will be disconnected from the opamp and the feedback will be terminated. This operation causes a charge accumulation at the input node of the opamp. To avoid the open-loop operation which will drive the output of the opamp to the saturation region, a fixed capacitor is connected in the feedback path of the opamp. Moreover, to ensure the gain to be the same for both clock phases, another fixed capacitor is connected in parallel to the input branch. The input–output relation of an amplifier with three inputs is given as

$$V_{out} = -\left(\frac{C_1}{C_F}V_1 + \frac{C_2}{C_F}V_2 + \frac{C_3}{C_F}V_3\right) \quad (26)$$

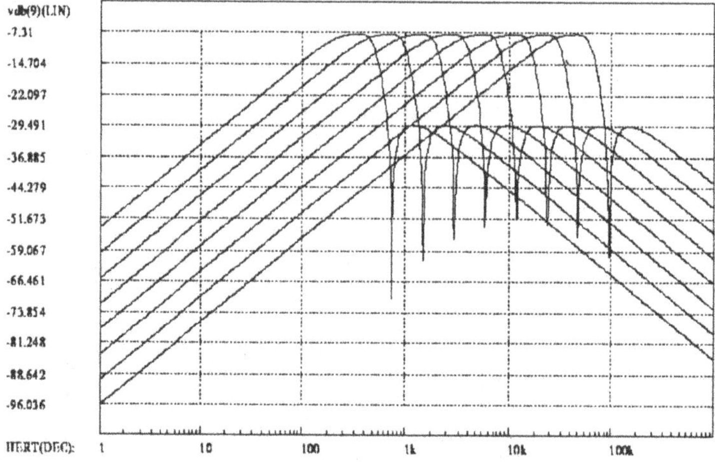

Fig. 16. Frequency response of the fourth-order band-pass filter with the clock frequency programmed from 2 kHz to 256 kHz.

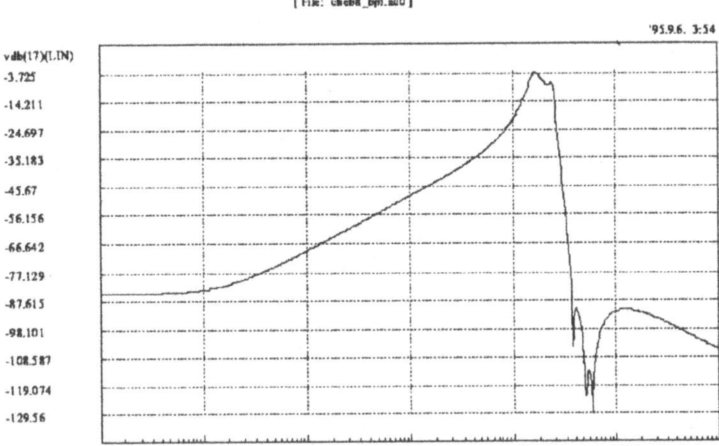

Fig. 17. Frequency response of the eighth-order Chebyshev band-pass filter.

If the capacitor values are chosen as binary increments of a unit capacitor C, a multiplying digital-to-analog converter (MDAC) with the following design equation can be realized.

$$gain = -\frac{\sum_{j=1}^{n} v_j 2^{n-j} C}{C 2^n} = \sum_{j=1}^{n} v_j 2^{-j} \quad (27)$$

A single lossy integrator of any CAB is enough to embed this circuit. A three-bit MDAC transient simulation is shown in Fig. 21.

5.2. *Balanced Modulator Implementation*

A balanced modulator [6] is realized by modulating the input signal by a square-wave carrier signal. This is equivalent to altering the clock phases with the carrier frequency so that the integrator operates at the high values of the carrier signal as an inverting integrator and at the low values of the carrier signal as a non-inverting integrator. The modulator is also embedded by configuring one of the CABs as a lossy integrator and disconnecting the rest. The transient

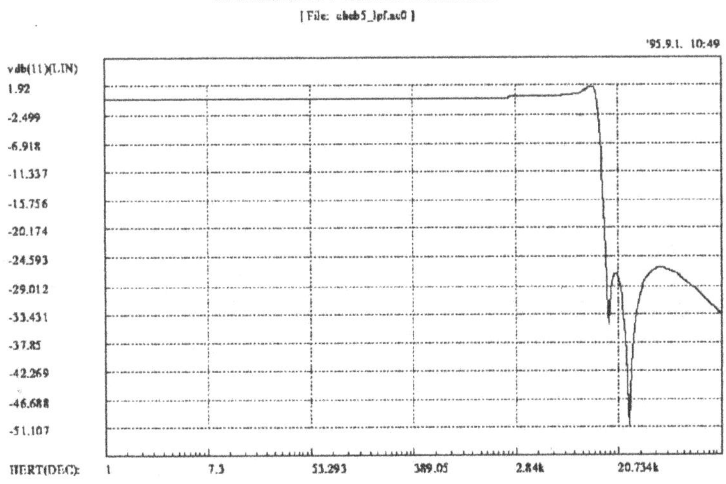

Fig. 18. Frequency response of the fifth-order Chebyshev low-pass filter.

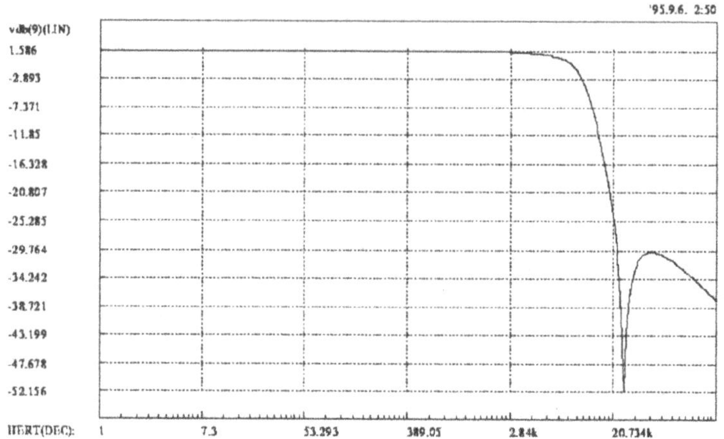

Fig. 19. Frequency response of the third-order (configured from fifth-order) low-pass filter.

simulation of the embedded balanced modulator is shown in Fig. 22.

5.3. Signal Generator Realization

By applying a square-wave signal to a high-Q band-pass circuit, a sine-wave can be generated. A quadrature sinusoidal generator is implemented using a single CAB by configuring the lossy integrator to a lossless integrator. The quadrature sinusoidal generator [11] with the design equations given as

$$C_1 = \frac{\pi}{4}\frac{V_{sin}}{V_{SQ}}\frac{1}{Q}C_{int1} \tag{28}$$

$$C_2 = \frac{1}{Q}C_{int1} \tag{29}$$

$$C_3 = 2\sin\left(\frac{\omega_0 T}{2}\right)C_{int1} \tag{30}$$

$$C_4 = 2\sin\left(\frac{\omega_0 T}{2}\right)C_{int2} \tag{31}$$

has an output with 180° phase shift and another

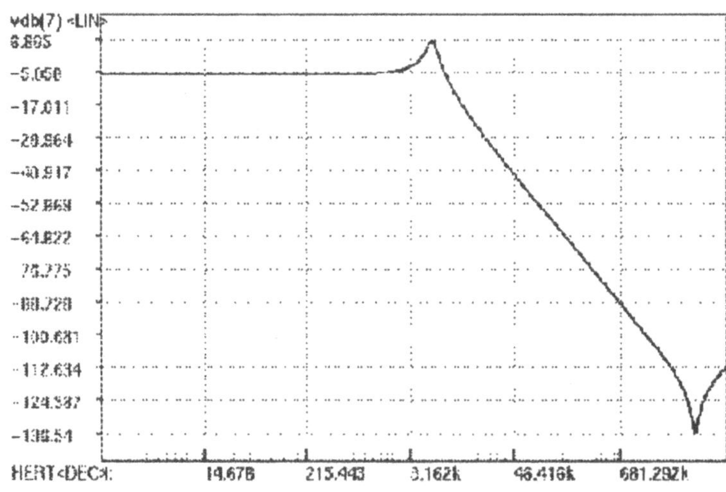

Fig. 20. Frequency response of the high-Q low-pass biquad circuit.

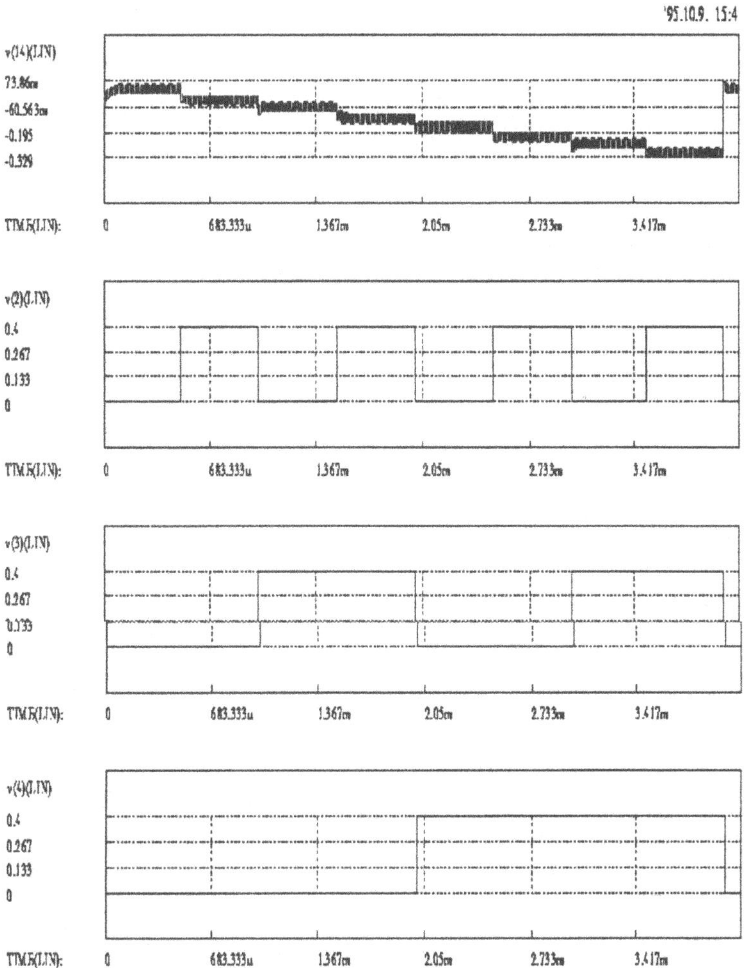

Fig. 21. 3-bit MDAC transient response.

sinusoidal output at the first opamp with 90° phase shift. The simulation result of the embedded signal generator is shown in Fig. 23.

6. Concluding Remarks

We have presented a voltage-mode switched-capacitor FPAA design. Various analog blocks are accomplished by utilizing the existing switches in the FPAA architecture, as well as clock phases and clock frequency for digital programming and topology configuration purposes. The FPAA architecture is used to design different types of analog filter structures, biquads, programmable amplifiers, modulators and signal generators via user-programmable digital control words. The proposed architecture provides easy and fast modifications to the desired specifications and/or applications. The functionality of the FPAA can be increased with use of more complicated CABs and digital control circuitry.

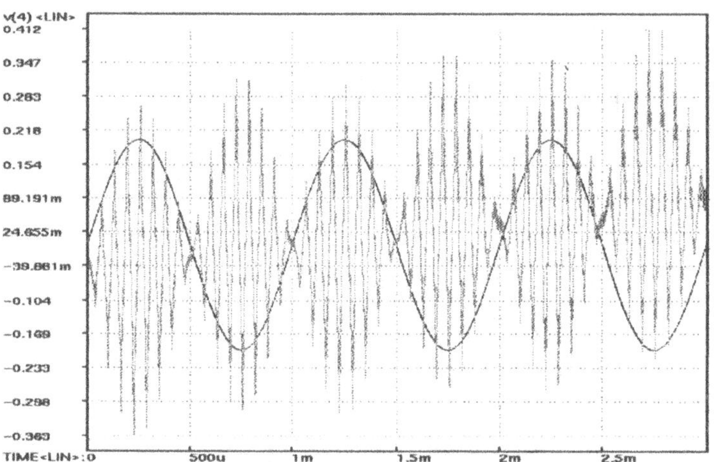

Fig. 22. Transient response of the balanced modulator to 1 kHz sinusoidal input signal.

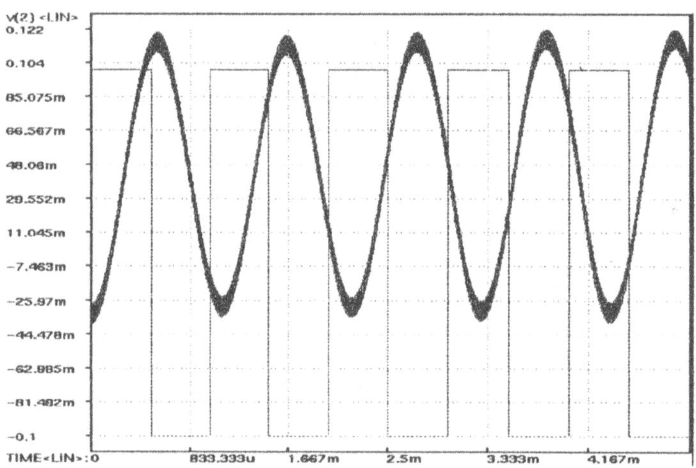

Fig. 23. Transient response of the quadrature sinusoidal generator.

Notes

1. In this paper the term programmability refers to changing any circuit parameter and configurability to changing the circuit structure or topology.
2. The term unswitched capacitor refers to a capacitor which is placed between programming switches but not between non-overlapping clock signal applied MOS transistors.

References

1. E. K. F. Lee and P. G. Gulak, "A CMOS Field-Programmable Analog Array." *IEEE J. Solid-State Circuits*, pp. 1860–1867, Dec. 1991.
2. E. K. F. Lee and P. G. Gulak, "Field Programmable Analogue Array Based on MOSFET Transconductances." *Electronics Letters*, pp. 28–29, Jan. 1992.
3. E. Pierzchala, M. A. Perkowski, P. V. Halen, and R. Schaumann, "Current-Mode Amplifier/Integrator Design for a Field-Programmable Analog Array." *IEEE International Solid-State Circuits Conference*, pp. 196–197, 1995.
4. H. Kutuk, "A Field-Programmable Analog Array (FPAA) Design Using Switched Capacitor Techniques." *M.S. thesis*, Dept. Elec. and Comput. Eng., University of Illinois at Urbana-Champaign, Urbana, 1996.
5. H. Kutuk and S. M. Kang, "A Field Programmable Analog Array (FPAA) Using Switched Capacitor Techniques." *1996 ISCAS Proceedings* 4, pp. 410–414, May 1996.

6. R. Gregorian and G. C. Temes, *Analog MOS Integrated Circuits For Signal Processing*. John Wiley and Sons, 1986.
7. D. Allstot, R. Brodersen, and P. Gray, "An Electrically Programmable Analog NMOS Second Order Filter." *1979 ISSCC Dig. Tech. Pap.*, pp. 76–88, Feb. 1980.
8. K. Martin, "Improved Circuits For The Realization Of Switched-Capacitor Filters." *IEEE Transactions on Circuits and Systems*, pp. 237–244, Apr. 1980.
9. P. R. Gray and R. G. Meyer, "MOS Operational Amplifier Design A Tutorial Overview." *IEEE J. Solid-State Circuits*, pp. 969–982, Dec. 1982.
10. B. D. Datar and A. S. Sedra, "Exact Design of Strays-Insensitive Switched-Capacitor Ladder Filters." *IEEE Transactions on Circuits and Systems*, pp. 888–898, Dec. 1983.
11. K. Martin and A. S. Sedra, "Switched-capacitor building blocks for adaptive systems." *IEEE Transactions on Circuits and Systems*, pp. 576–584, June 1981.

Haydar Kutuk was born in Izmir, Turkey. He received the B.S. degree from Istanbul Technical University in 1992 and the M.S. degree from University of Illinois at Urbana-Champaign in 1996, both in electrical engineering. He is currently working towards the Ph.D. degree in electrical engineering at the same university. His Ph.D. research is focused on interconnect simulation techniques for high speed timing simulators. His research interests include computer-aided design of VLSI circuits, analog and digital VLSI circuit design for signal processing and digital and neural signal processing.

Sung-Mo (Steve) Kang received the Ph.D. degree in Electrical Engineering from the University of California at Berkeley. He has worked on CMOS VLSI design at AT&T Bell Laboratories at Murray Hill, N.J. as Supervisor and Member of Technical Staff of High-end CMOS VLSI Microprocessor Design. Currently, he is Professor and Head of the Department of Electrical and Computer Engineering, Charles Marshall University Scholar, and Research Professor of Beckman Institute for Advanced Science and Technology, and the Coordinated Science Laboratory at the University of Illinois at Urbana-Champaign. His research interests include computer-aided design of VLSI circuits and systems, design optimization for performance, reliability and manufacturability, modeling and simulation of semiconductor devices and circuits, optoelectronic integrated circuits and fully optical networks. He was the Founding Editor-in-Chief of the IEEE Transactions on Very Large Scale Integration (VLSI) Systems and has served on editorial boards of several IEEE and international journals. He has received a Humboldt Research Award for Senior US Scientists, the IEEE Graduate Teaching Technical Field Award, IEEE Circuits and Systems Society Meritorious Service Award, SRC Inventor Recognition Awards, IEEE CAS Darlington Prize Paper Award and other best paper awards and co-authored five patents, over 250 papers, and six books, Design Automation For Timing-Driven Layout Synthesis, Hot-Carrier Reliability of MOS VLSI Circuits, Physical Design for Multichip Modules, and Modeling of Electrical Overstress in Integrated Circuits from Kluwer Academic Publishers, CMOS Digital Integrated Circuits: Design and Analysis from McGraw-Hill, and Computer-Aided Design of Optoelectronic Integrated Circuits and Systems from Prentice-Hall.

Analog Integrated Circuits and Signal Processing, 17, 67–89 (1998)

DPAD2—A Field Programmable Analog Array

ADRIAN BRATT AND IAN MACBETH

Motorola Programmable Technologies Centre, Sherwood House, Gadbrook Park, Northwich, Cheshire CW9 7TN, England

Received June 28, 1996; Accepted April 23, 1997

Abstract. DPAD2 is a Field Programmable Analog Array (FPAA) based on CMOS switched capacitor technology. This paper describes the major design decisions that went into creating DPAD2 with respect to the ultimate goal of the work, being a mixed signal field programmable silicon solution.

Two major compromises exist in the design of an FPAA, one between flexibility and performance, the other between functionality and die size; DPAD2 overcomes the first with a novel field programmable hierarchic routing scheme and the second by careful analysis of many disparate designs to arrive at a best compromise solution. Results from prototype silicon are presented where a single analog cell is reconfigured to perform a number of different analog signal processing functions. Bandwidth of the DPAD2 device is 500 kHz and the SNR is typically 60 dB, although both are application dependent.

Introduction of the FPAA now enables a designer to have working silicon within one day, by a simple configuration of the silicon chip via a PC parallel interface. Software libraries of analog circuits are provided and allow very rapid creation of large and complex analog circuits.

Key Words: analog field array, CMOS

1. Introduction

Field Programmable Gate Arrays (FPGAs) are currently enjoying considerable success because of the ease of configuration and reconfiguration, even though there is usually some loss of performance when compared to an ASIC. It is also apparent that many silicon designs are now converging towards single chip solutions which are frequently mixed-mode. Combining these two trends it is obvious that there is a large emerging market for field programmable parts with mixed-mode capability. At present, the missing part of the solution is an FPAA and such an array is described in this paper. From the outset this FPAA was tailored towards inclusion in a Field Programmable Mixed-signal Array (FPMA). In this paper we discuss an FPAA that has been designed to be process compatible with an existing FPGA technology (the Motorola MPA1000 family) and we present some results from the analog array.

Contact Author, Dr. Adrian Bratt, Motorola Programmable Technologies Centre, Sherwood House, Gadbrook Park, Northwich, Cheshire CW9 7TN, England, E-mail: adrian@pmel.com.

Functionally, a successful FPAA must exhibit a number of features:

1. Random routing between cells must be possible to allow a true array architecture. In particular, choice of routing (often automatically performed) must not cause a significant change in the circuit transfer function.
2. Interfacing between digital and analog domains must occur naturally, without large amounts of complex interface circuitry.
3. Analog transfer functions must be easily calculated and implemented.
4. A large base of existing circuit techniques must exist to aid the speedy implementation of diverse functions.
5. Transfer functions must be accurate and show little drift or initial tolerance.

In light of these requirements, it is possible to make an assessment of the present FPAA designs that are available from both industry and academia. This is done in terms of their suitability as FPMA designs, although this was not the direct intention of the designers in some cases.

The first entrant to the market were the 5010 and

5030 devices from IMP [1]. Both these devices are aimed at data acquisition and essentially have a linear data flow from analog to digital domains. These devices are aimed at a specific task and most certainly cannot be regarded as arrays. The functional blocks within them are capable of some parametric programming but have functions that are inflexible. In considering the list of FPAA specifications above, the IMP devices fail to satisfy points one and four.

The University of Toronto has published an analog array in [2]. The base technology of this circuit is transconductors, programmable resistors and programmable capacitors which allow filter like structures to be implemented. A central cross-routing box allows quite a high degree of flexibility in interconnecting the components, but the chip is limited in its choice of base technology to implementing gm-C filter like structures. These are not accurate circuits without trimming, because both the transconductor gm value and capacitance each show uncorrelated manufacturing tolerances of 20% or so. Temperature performance of such circuits is also generally poor because of poor matching of temperature coefficients. Consequently, this design fails to satisfy point five in the requirements list above although it does satisfy the other four points to some degree.

TRAC [3] is a field programmable device recently announced by Zetex. This is an array of cells which operate in the log anti-log domain, using the p-n junction of a bipolar transistor in the feedback loop of an operational amplifier. For mathematical functions this device offers some attractive features, for example multiplication may be done using the log, add, anti-log technique. Filter structures are implemented using off chip components, which immediately adds manufacturing error tolerance into the transfer function, from both the capacitors and the transconductor elements. Again, temperature tolerance is poor because of the poor matching of capacitors and transconductors. This design fails to meet point five in the list above, it also fails on point two in that there is no obvious way to add digital interfacing without major interface circuitry. The amount of routing possible within the TRAC design is quite limited, unless a significant number of cells are dedicated to buffering, and the design also fails to meet point one as a result.

FIPSOC [4] is a device recently announced from a consortium of universities and an industrial partner. The analog part of this mixed-mode chip is a coarsely granular structure which comprises gain blocks, filters, analog multiplexors and ADC/DAC cells. This choice of cells does represent a wide selection of the analog designer's tool kit but the user is essentially constrained to use the cells provided by the designers. Inclusion of full custom cells does lead to very efficient implementations of the blocks provided, but it does limit the diversity of functions that the end designer may create.

A review of the present FPAA devices must thus conclude by saying that some come close to achieving the goal of an FPAA but all fail in some respect. In considering what might succeed it is now necessary to review the available technologies.

There exist a number of circuit design techniques which may be employed as the base technology for an FPAA. Examples are switched current [5], transconductor-capacitor [6] and MOSFET-capacitor [7], but the only one which satisfactorily addresses the above criteria is switched-capacitor (SC) [8]. Specifically to address the five points above it can be observed that:

1. Arbitrarily routed circuits must necessarily be parasitic insensitive circuits [9] because of the widely different loads presented by different routing paths. SC architectures exist which are insensitive to grounded parasitic capacitors, this principle is demonstrated by the SC integrator shown in Fig. 1. Nodes within this circuit are either virtual earth nodes or are driven by strongly buffered voltages. Parasitic capacitances to ground on virtual earth nodes hold an almost constant charge and so have very little effect. All other nodes are driven by strongly buffered voltages and are easily charged and discharged without charge sharing errors. The two clock phases for this circuit are denoted ϕ_1 and ϕ_2 and are arranged in a non-overlapping sequence, such that a switch controlled by ϕ_1 is never turned on at the same time as a switch controlled by phase ϕ_2. The relationship of such clock phases to each other is shown diagramatically in Fig. 2.
2. SC circuits rely on switches for their function. Interfacing to digital control signals occurs very naturally as a result. Programming of capacitor values is also accomplished very simply, as in Fig. 3, where a number of capacitors are combined in parallel. Simply by opening or closing the pass transistors (under digital control) it is possible to vary the effective capacitance of the array between the IN and OUT terminals.

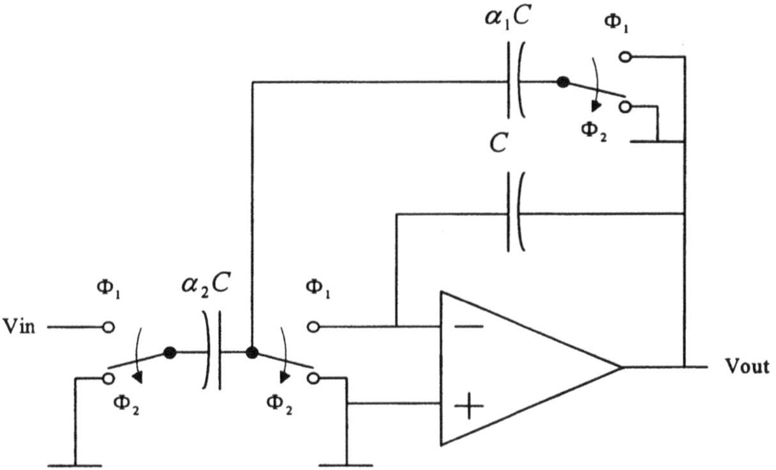

Fig. 1. A switched capacitor lossy integrator.

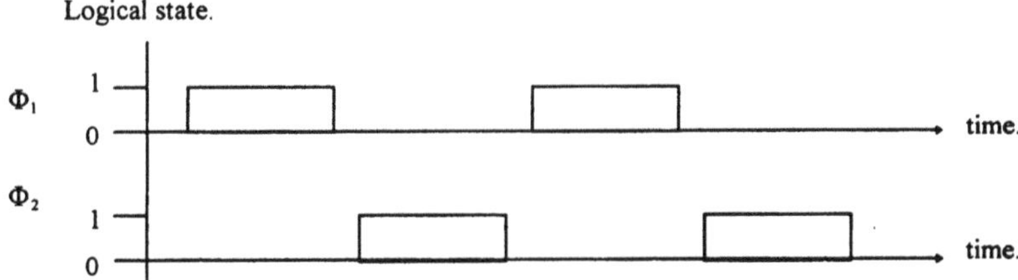

Fig. 2. Two-phase non overlapping clocks.

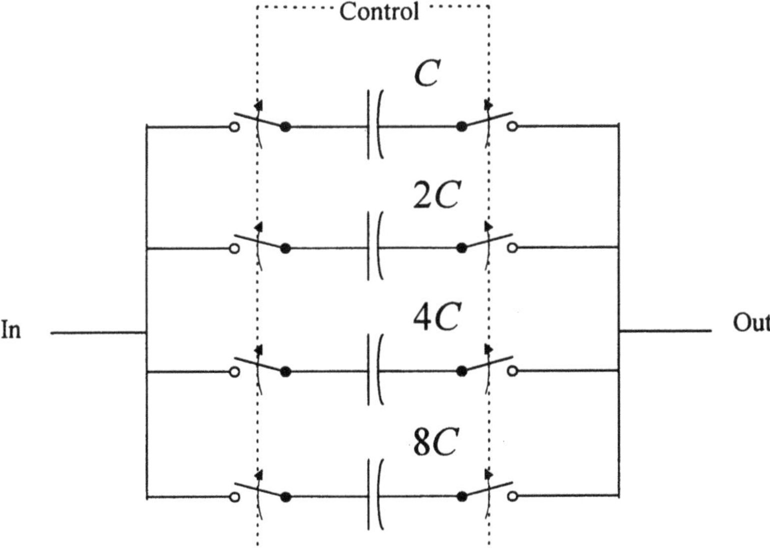

Fig. 3. A programmable capacitor array.

3. Transfer functions are easily calculated (often orthogonally) by a combination of capacitor ratios and clock period.
4. SC circuits have been in existence since the mid 1970s [10] and a huge database of design techniques is available, many of these being parasitic insensitive.
5. SC circuits usually have a well defined transfer function. For example, the transfer function of the circuit in Fig. 1 is given by,

$$\frac{V_{OUT}}{V_{IN}}(z) = \frac{\frac{\alpha_2}{1+\alpha_1}}{1 - \frac{1}{1+\alpha_1} z^{-1}}$$

where z^{-1} is equivalent to a delay of one clock period (a very precisely defined value) and the value of α_1 is the ratio C_1/C (similarly α_2 is equal to C_2/C). On a good quality analog fabrication process α_1 and α_2 may be controlled to an accuracy of 0.1% or so. The resulting transfer function is well defined and shows little drift or initial tolerance. Temperature tracking of the two capacitors is usually very close, making the α terms insensitive to temperature.

SC technology is clearly the optimum choice for an FPMA but, even so, major technical difficulties remain if the FPAA is to succeed. The major technical barrier to successful implementation of an FPAA is, indirectly, the operational amplifier unit cell of analog design; Such a large building block means that the FPAA structure is necessarily coarse, as a consequence implementation of resources is a very critical task. With too few operational amplifiers, an FPAA would not be capable of meaningful functions. Conversely, features that are seldom used add considerably to the die size. To address the functionality/die size trade-off a large number of SC designs were created using the Cadence design suite [11], the generic groups for this process are listed in appendix A. A utility was then written to search the data-base and analyze the designs in terms of the number of operational amplifiers used, the number of input branches and the spread of capacitance values within a circuit. From this data the unit cell of the DPAD2 array was designed as a best guess at the most efficient compromise solution. The unit cell of the DPAD2 array is described further in Section 3.

The functionality/size trade-off is very important, but the trade-off between flexibility and performance is equally important. For the SC technology used for DPAD2, extra flexibility can of course be obtained by creating extra switching options between nodes. The price paid is that each switch increases loading of the operational amplifiers, adds series resistance to the signal paths, and most importantly each switch leaks charge from the nodes connected to it [12]. This charge leakage is primarily associated with the drain or source implant of a MOSFET and exhibits the characteristic leakage current of a reverse biased diode, as in Fig. 4. If such a switch is connected to the summing node of an integrator then the leakage charge is integrated, leading to gradual corruption of

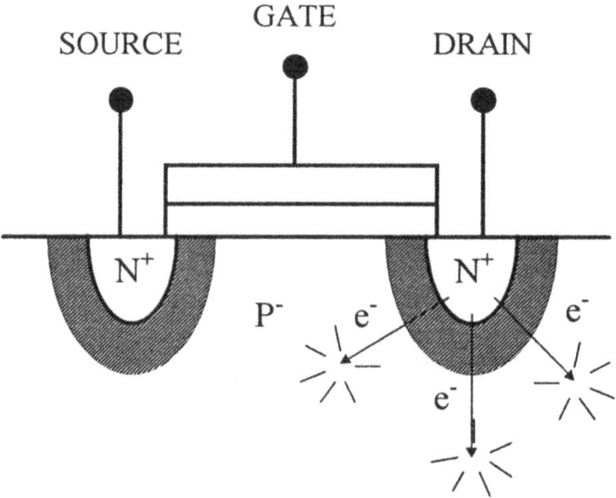

Fig. 4. Charge leakage from a NMOS transistor drain.

the signal. Implementing an excessive number of switches is thus to be avoided to maintain performance, but sufficient must be present to allow arbitrary routing between cells in the array, and between nodes in a core cell. The solution adopted for DPAD2 was the use of a hierarchic routing scheme, with routing at both a local and at a global level within the array. This is further described in Sections 2 and 3 of this paper.

Programming software for the DPAD2 device is described in Section 4 and some example circuits are shown configured onto the DPAD2 FPAA in Section 5 using this software tool. Results from prototype silicon are presented in Section 6. Section 7 concludes the report and discusses future work.

2. System Level Resources

2.1. *Array Overview*

Fig. 5 shows a schematic view of the DPAD2 FPAA at the highest hierarchical level. The core of the chip is occupied by an array of 20 (five by four) core cells and their contents are described further in Section 3. Each of these cells is identical and each may communicate with any other in the array via field programmable interconnect. It is a vital feature of an FPAA that the user must not be confined to a rigid signal flow, since analog design is fundamentally a very creative process.

A total of 13 I/O cells are provided around the periphery of the chip and each contains a single operational amplifier, which may drive onto or off the chip. These cells may be used in conjunction with passive external components to perform as anti-alias [13] and smoothing filters when necessary. Peripheral cells may also be used as unity gain buffers if desired or it is possible to drive direct onto/off the array with no buffering.

The upper side of the die in Fig. 5 has a large shift register for configuration—this is loaded in serial and written in parallel to other blocks of SRAM in the array. Above this shift register are digital I/O pads for interfacing. On the present array this digital interface is pinned out for direct external access, but on the mixed array this is the point at which the digital/analog interface will exist. A global clocking scheme is provided (not shown) which associates one of four clock signals to each cell of the array, this maintains synchronization of the cells during signal processing. In total there are four clock input pads and each of these may be driven at a different frequency and connected to arbitrary cells within the array. Using different clock frequencies allows decimating and interpolating functions to be implemented [14] and

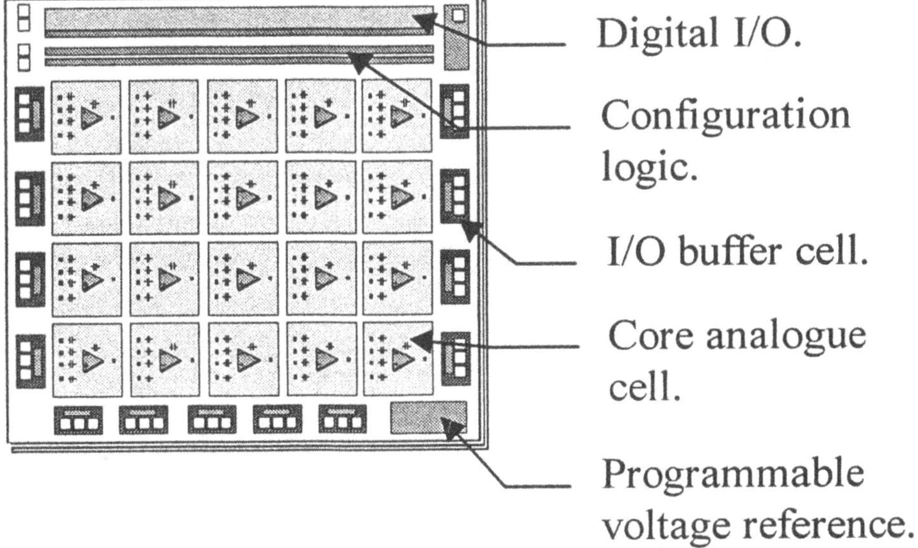

Fig. 5. Array architecture overview.

allows parallel signal paths within the array operating at different clock frequencies. For the mixed signal array it is anticipated that the clocking signals will be controlled from the FPGA section, allowing complex stop-starting and gating of clock signals under specific conditions defined by the user.

The lower right of Fig. 5 shows the location of the programmable voltage reference. This is field programmable and is based on a bandgap reference. It has eight bit resolution and provides a reference voltage for all the operational amplifiers within the array. The voltage at the virtual earth of an operational amplifier is set by this voltage.

2.2. *Global Interconnect*

A two wire bus runs horizontally, and another runs vertically, between each row and column of cells, as shown in Fig. 6. These are referred to as global routing busses and their function is to carry signals up to the whole width of the array. They allow routing between arbitrary cells within the array, with a clean partition between signals at a more local level. The elegance of the DPAD2 hierarchic routing can be seen by considering a non-hierarchic routing scheme that allowed arbitrary single connections between 20 outputs and 80 possible input points (assuming 4 input branches per cell). This would need to have 80 switches per output in parallel, making a total of 1600 switches. Each of these switches would need a RAM cell to control its state and there would need to be a bus structure that was 1600 tracks wide. With the DPAD2 global routing scheme each operational amplifier has only four outputs onto the global interconnect, allowing connection to the four global nets proximal to each analog cell. At each intersection of the vertical and horizontal global busses a switch matrix is provided, to allow dog-leg routing between cells across the array that are not placed immediately above/below or left/right of each other.

Two other busses are provided at the array level, the data transfer bus (DTB) which carries data for configuration of cells and the transfer control bus (TCB) which is used for latching of data from the data transfer bus and for various house keeping functions. Dynamic transfer of configuration values can occur via the DTB while the DPAD2 array is processing a signal. This allows functions such as dynamic gain control and adaptive filtering to operate under external digital control.

3. Local Level Resources

3.1. *Local Routing*

In addition to long distance signals routed through the array, there are signals that only travel a short distance, the input nets and feedback network of

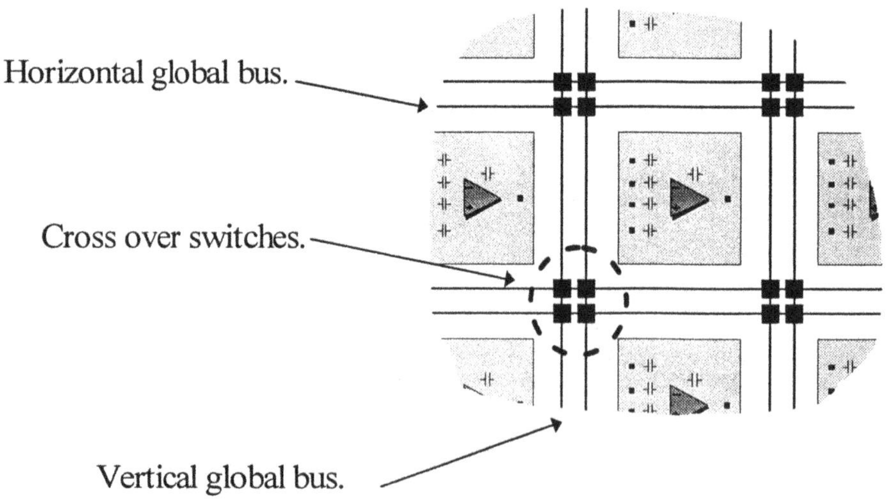

Fig. 6. Global busses and cross over switches.

Fig. 1 being a prime example. A dedicated layer of interconnect (local interconnect) has been implemented for such signal paths. Fig. 7 shows the fan-in arrangement of local interconnect for a typical cell in the center of the array. Each cell has four inputs, each of which has local interconnect dedicated to a set of surrounding cells, denoted by the letters ''B'', ''C'' and ''D''. Input ''A'' is dedicated to the output of its own cell, this special feature is implemented because an operational amplifier almost always has at least one feedback connection of this type. The ''D'' input is also special, in that one of its branches comes from a cell two to the left of the present cell. This feature was implemented to facilitate mapping of designs onto the array from the usual design flow from left to right on paper.

At the point where the various local interconnect wires enter a cell there is one pass transistor per input. To select the appropriate input to a cell (say from one of the three possible ''D'' cells) it is a simple matter to activate the appropriate pass transistor at the dedicated ''D'' input. The parallel signals feeding to the input multiplexors are always driven by the appropriate operational amplifier output, thus only one pass transistor is present in the local interconnect. Thus each cell has the capability to accept a signal from the nine cells around it without the need to use global interconnect at all.

At the edge of the array the peripheral I/O cells act as members of the array where possible. Routing on and off the array is thus accomplished smoothly without special interface circuits. Since most signals are routed locally this frees the global interconnect to route long distance signals. In the event that local interconnect becomes fully utilized, it is possible to route locally using the global busses, since connections are provided to route from various points on the local interconnect to the global busses and vice-versa. Obviously this is not a preferred option because it represents a less efficient routing scheme.

3.2. Core Cell Functions

Fig. 8 shows a schematic diagram of the core analog cell. Central to the scheme is an operational amplifier which provides a virtual earth for the switched capacitors, it also buffers the output voltage of the circuit. The local interconnect signal paths are shown entering on the left of Fig. 8 and the special connectivity of the ''A'' input is also shown. Within the core cell there exist a number of internal busses that may be used to multiplex signals from one input group to another. Generally this is not necessary, because the Motorola FPAA layout software has the ability to re-arrange the input branches from the user specified positions. For example, a cell which is in the ''C'' position in Fig. 7 is best routed to the central cell

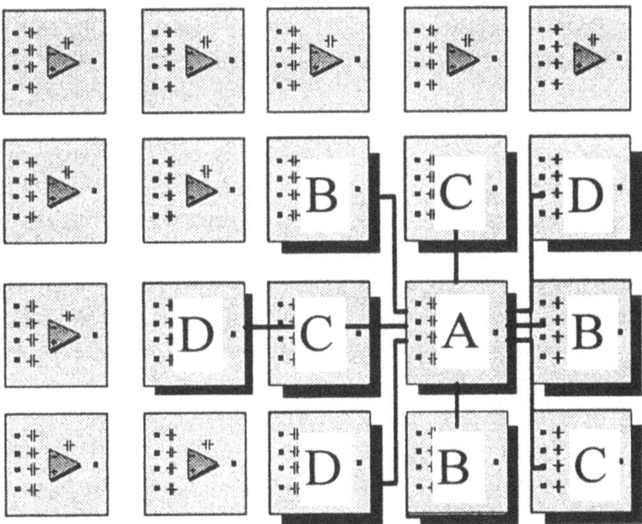

Fig. 7. Local interconnect fan-in.

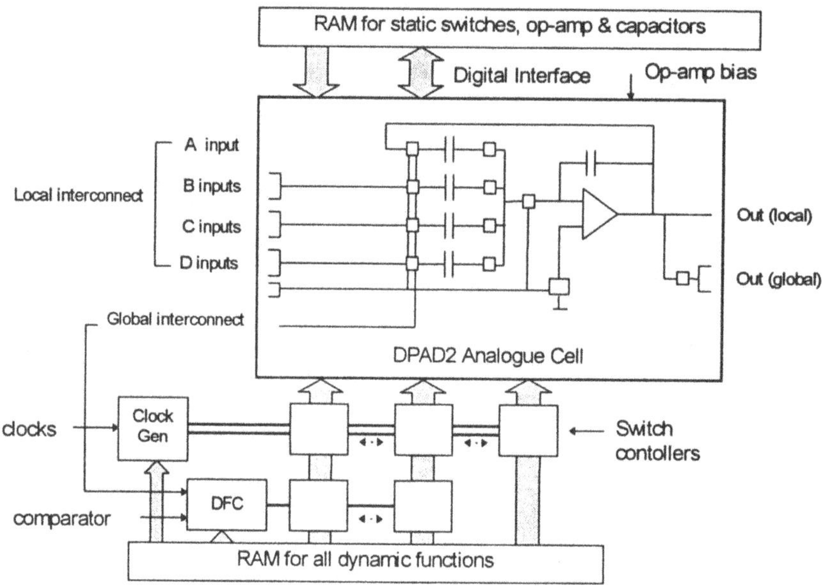

Fig. 8. Schematic view of a core cell.

over local interconnect that enters the "C" input. If the user were to connect a "C" cell to a "D" input then the software would alter this arrangement to enter the cell on the "C" input.

Internal busses of the cell allow many and varied routes to be utilized between capacitor and operational amplifier nodes. Use of parasitic insensitive technology is demonstrated in its full strength here, because each of these routes has a different parasitic capacitance associated with it. A circuit which is parasitic insensitive shows almost no change in transfer function for an arbitrary route, while the performance of a parasitic sensitive circuit would be very difficult to predict.

Fig. 8 shows switch controller blocks operating on the clock signal. These controllers allow many different internal cell configurations which would not be possible by directly connecting the clock to the switches, as is usually done in SC circuits. The switch decoders perform a decode of the clock signal which they are fed with, a switch may close on clock high, on clock low or be permanently open or closed. Signal dependent functions are also possible via the Dynamic Function Control (DFC) block which allows dynamic control of switch function while the circuit is operating. It takes the output of a comparator (part of the core cell but not shown) and is able to alter the phase and timing of various switches dependent on the comparator output value. Such a function readily finds use in the implementation of a rectifier, as is demonstrated later.

Two SRAM banks are shown in Fig. 8. The lower SRAM bank holds the data for the functions that are termed dynamic functions, i.e. the switch decoders which decode the master clock within the cell and feed the result to the dynamic switches. The upper SRAM bank holds the configuration data for the capacitor values, the static switch configurations and the bias condition of the operational amplifier. These are termed static values which are not usually changed after a full device configuration has taken place, although it is possible to dynamically change capacitor values if they are marked as such at configuration time.

As well as the usual SC functions, the core cell is able to do numerous other functions. A route exists direct to the non-inverting terminal of the operational amplifier and this may be used for a unity gain buffer function. Removing the operational amplifier feedback makes the circuit perform as a comparator. The possible permutations of the core cell are very large indeed, this flexibility is considered to be essential for a FPAA as described previously. To emphasize the great flexibility of the DPAD2 cell it is possible to

Table 1. One-cell macro functions.

Cell name	Parameters
Cosine filter	Gain, pole frequency
Gain	Gain
Polarity modulator	Gain, modulating signal
Rectifier	Gain, full/have wave, inverting/ non-inverting
Sample & Hold	Gain, CLK or CLKBAR
Track & Hold	Gain, gated or continuous
Schmitt trigger	Hysteresis
Summing stage	Input weight for 4 inputs, scaling, autozero
Ladder segment	Inductive or capacitative value
Ladder I/O termination	Equivalent resistance

list the present one-cell macro functions and the parameters associated with each of these, as in Table 1. These functions represent the present list and in no way represent a fixed set, the ingenuity of the designer will allow many more to be created.

4. Software

Motorola has developed a CAD system for the DPAD2 device called August. It is based on existing FPGA tools but is heavily modified to address analog applications. It allows the graphical entry of designs (either directly or following an EDIF net-list import) and can perform auto-routing of nets following a manual placement of components and higher level functions. The design flow is shown in Fig. 9. Initially, the design is entered as an EDIF netlist from a third party tool or as a manual layout. For the latter, the user progresses directly to the interactive layout system. Once in the interactive system, it is possible to use macro-blocks (for the less experienced user) or directly access primitive elements such as the capacitors and switches, for the expert. A mixture of the two design styles is also possible. Once placement and parameterization has finished the tool will perform auto-routing based on a minimum cost algorithm. Once routing is completed the user leaves

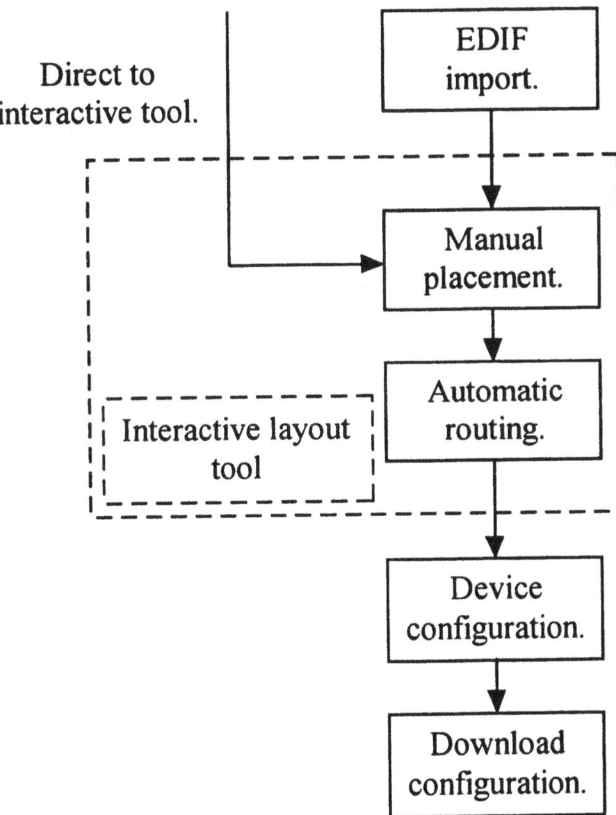

Fig. 9. August design flow.

the interactive tool and streams the data out as a configuration file. This file may be directly loaded into the DPAD2 chip via a driver program to perform the desired function. It is also possible to program an EEPROM and then use this to load a configuration into the DPAD2 chip.

5. Example Configurations

The great flexibility of the DPAD2 array allows a large number of circuit configurations to be implemented. This section demonstrates a small sample of these. Fig. 10 shows a schematic view of a cosine filter [15], a decimating filter which runs off both the even and odd phases of the same clock. Its implementation on the DPAD2 array using August is shown in Fig. 11. The switch designations SA to SE may be traced between the two diagrams as may the capacitor designations CA to CD. An unused core cell is shown at the base of Fig. 11. The cosine filter shown is driven from one of the clock drivers—there are three more drivers that are not utilized in this design. These may be used to provide other clocks of arbitrary frequency and duty cycle for other parts of the circuit. Generally speaking the whole circuit will often run from one clock—and one driver may be used for all cells in such a case. The analog input and output buffers on the left side of Fig. 11 are programmable for direction of signal flow, if desired the input and output pins may be exchanged under software control.

A function which involves the dynamic control of switches is a rectifier circuit [16], shown in Fig. 12. The dynamic function control (DFC) unit is capable of altering switch phases in various ways as a function of a comparison based on the input voltage and a reference voltage. Here the overall circuit function is periodically inverting then non-inverting so that the output voltage polarity is always positive. A configuration of this circuit onto the DPAD2 array is shown in Fig. 13 and again the switch and capacitor designations may be traced between Fig. 12 and Fig. 13. It is a simple matter to alter the function of the DFC unit to make either a full wave rectifier or a half wave rectifier and either may be made inverting or non-inverting.

Implementation of a large or complex design requires the use of hierarchy to manage the design. DPAD2 provides this by using large macro cells which have flexible architecture and parameters. An example of a larger design is the PCM CODEC [17] shown committed to the array in Fig. 14. The dark boxes are macro cells, four second order filter blocks are shown as well as a cosine decimating filter at the upper left. A u-law DAC [18] is shown in full detail in the lower center and occupies three cells.

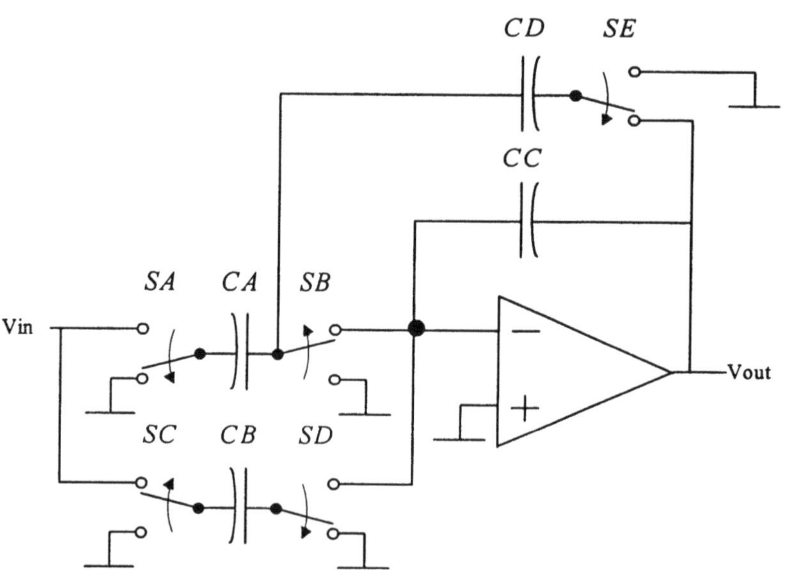

Fig. 10. Schematic of a decimating cosine filter.

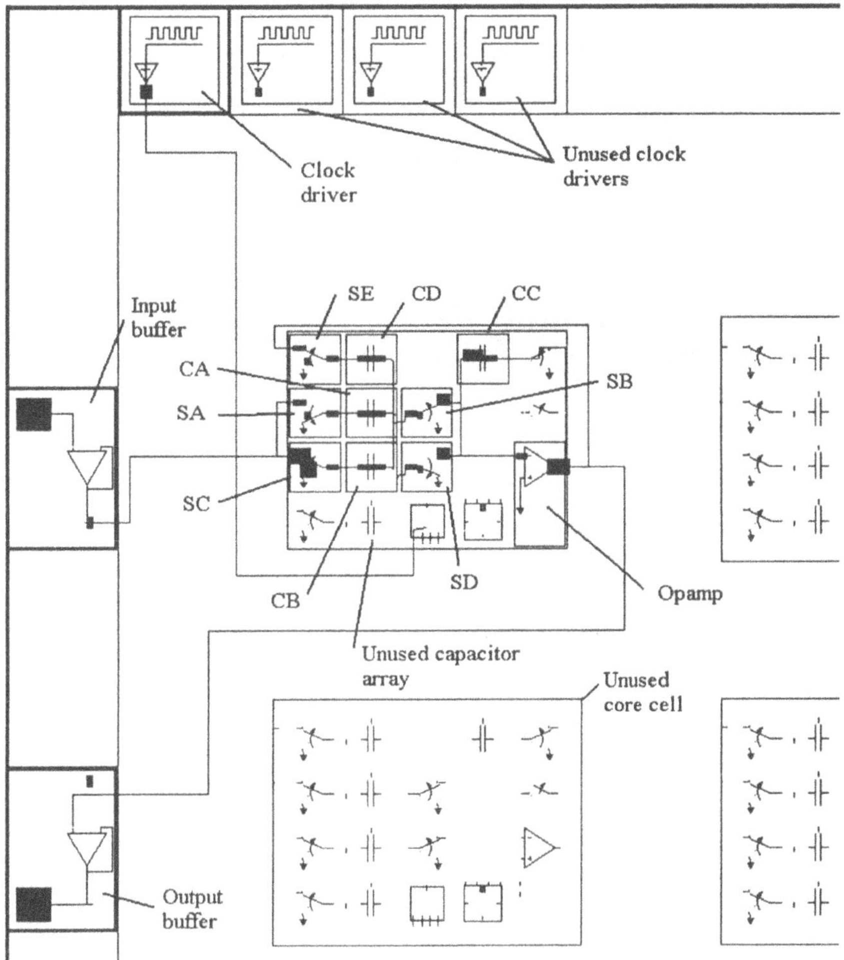

Fig. 11. Implementation of cosine filter on one analog core cell.

6. Results from Silicon

The major performance specifications for the DPAD2 array are shown in Table 2. As can be seen, the array is based on 8 bit programming for the capacitors within the unit cell and for the voltage reference support function. 1 MHz is the present maximum clock frequency although there are plans to improve this to increase the application scope of the device. At 1 MHz clocking with a 100:1 clock to signal ratio it is possible to process signals in the DC to 10 kHz frequency band. This encapsulates the voice band for many applications, although a higher frequency range

Table 2. DPAD2 performance figures.

Parameter	Value
Power supply	0,5 V
No. of analog cells	20
Analog i/o ports	13
Local interconnect connections per cell	9
Capacitor programming	8 bits
Voltage reference programming	8 bits
Max. clock frequency	1 MHz
Input signal range	0–5 v
THD per cell	0.4%
Typical supply current per cell	1.6 mA
Analog array configuration time	10 ms

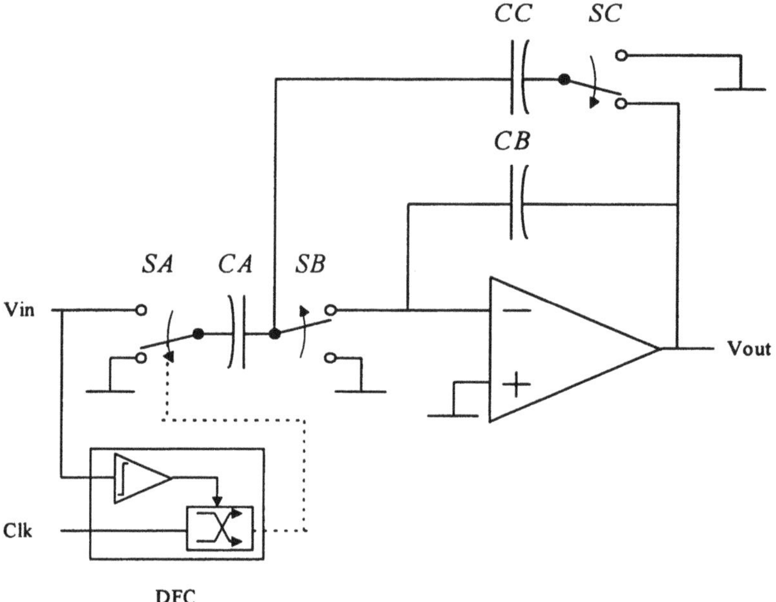

Fig. 12. Schematic view of a rectifier circuit.

would be necessary for high fidelity audio. The THD per cell is 0.4%, which is measured for a simple gain stage. This error is largely due to charge injection effects of the switches and is strongly a function of the circuit configuration. The power consumption of the array depends on how many cells are active. Those that are not active are automatically powered down, the active cells consume a static current of 1.6 mA each when fully powered up.

It should be noted here that all the results presented in this section are obtained from one analog cell which has been re-configured to perform all the functions. The schematic of a variable gain stage is shown in Fig. 15 and the resulting output waveforms for gains of -1, -2 and -3 are shown in Fig. 16, where the solid traces are three the output voltages and the broken trace is the input voltage.

Results for the full wave or half wave rectifier circuit of Fig. 12 are shown in Fig. 17 and Fig. 18 respectively. The full wave rectifier is shown operating in non-inverting mode and the half wave rectifier is also shown in non-inverting mode. On both plots the input voltage is shown as a broken line and the output as a solid line. These two results were obtained from the latest version of the FPAA, DPAD3, rather than DPAD2.

Fig. 19 shows the results from a relaxation oscillator [19], both the output of the comparator (the square-wave) and the internal ramping node (saw tooth) are shown.

The sample and hold circuit [20] and the Schmitt trigger [21] circuit are two sample circuits from appendix A. The schematics of these circuits are shown in Fig. 20 and Fig. 21 respectively. The associated circuits configured onto the DPAD2 array are shown in Fig. 22 and Fig. 23 respectively. Fig. 24 and Fig. 25 show the results of the two circuits in operation with a sine-wave input. The Schmitt trigger is essentially a comparator with adjustable hysteresis, in this example the hysteresis value was set to 1.0 V and the achieved value was 0.967 V, an error of 3%.

7. Conclusions

The DPAD2 architecture presented in this paper is a true analog array and allows a very large degree of

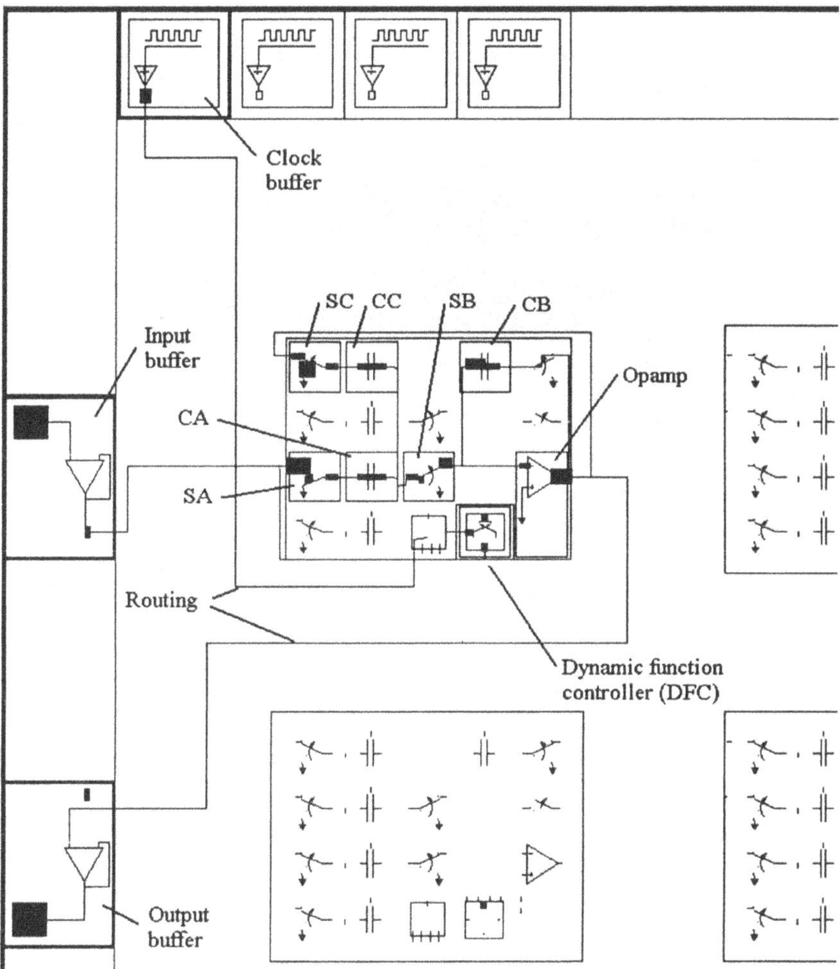

Fig. 13. Implementation of a rectifier circuit on one analog core cell.

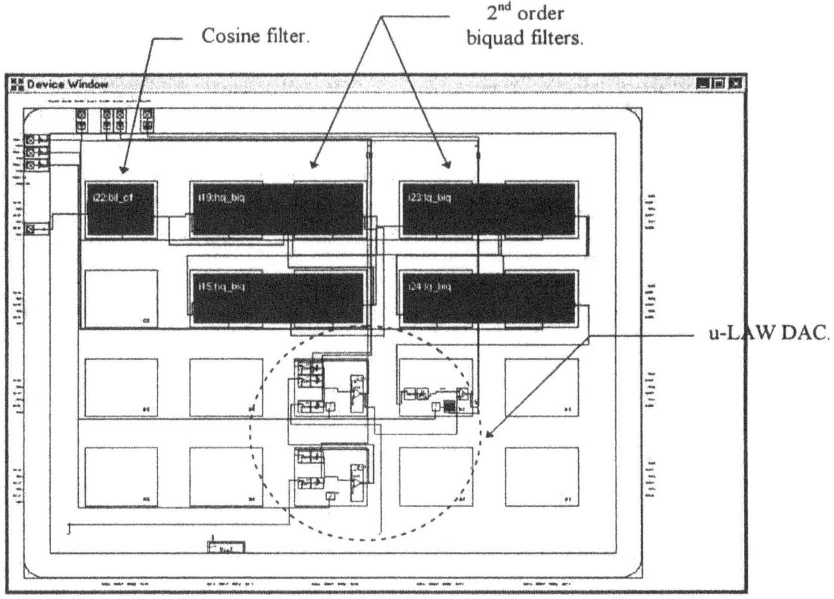

Fig. 14. PCM CODEC implementation.

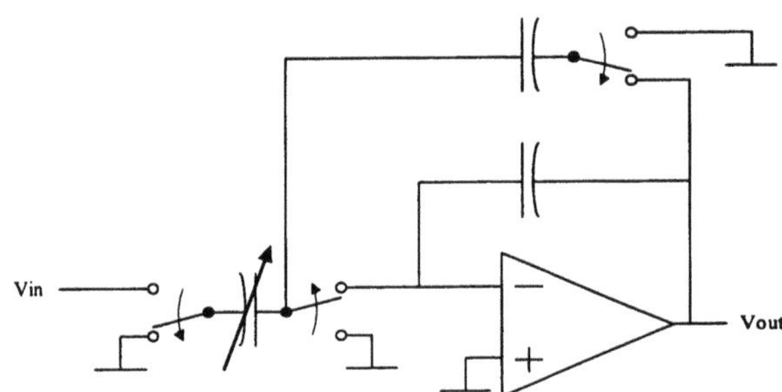

Fig. 15. Variable gain stage schematic.

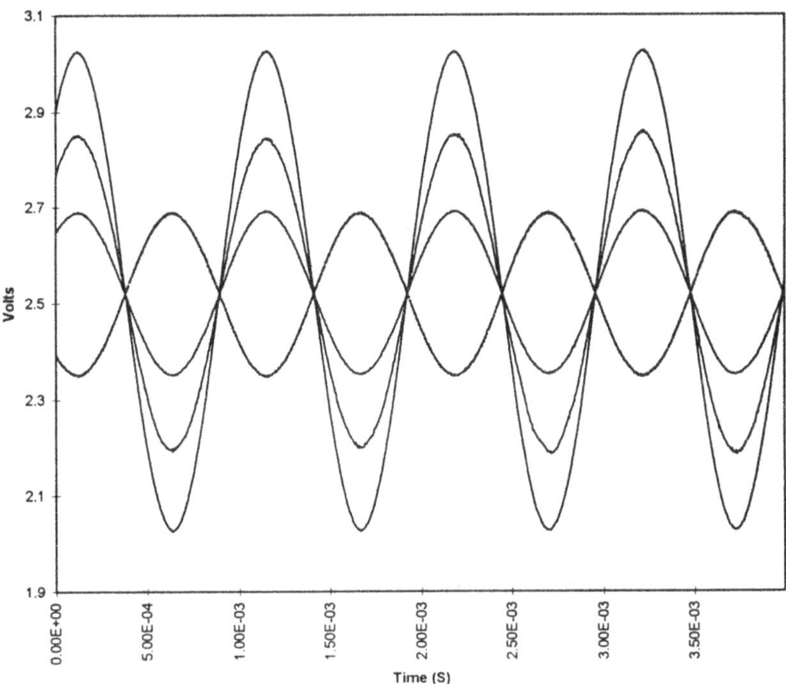

Fig. 16. Input and output waveforms of variable gain stage.

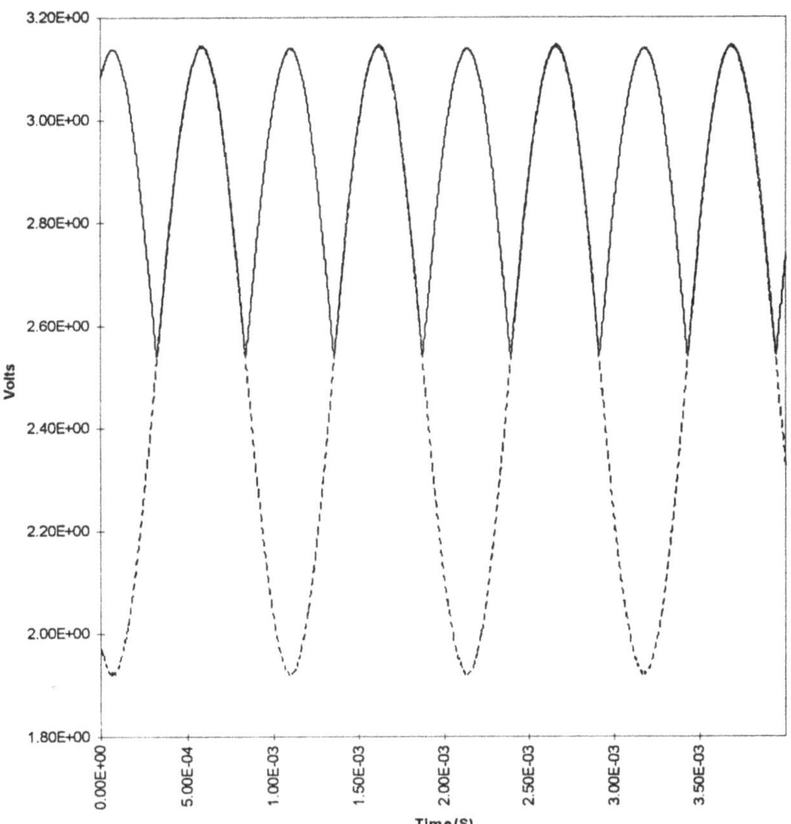

Fig. 17. Full wave rectifier input and output traces.

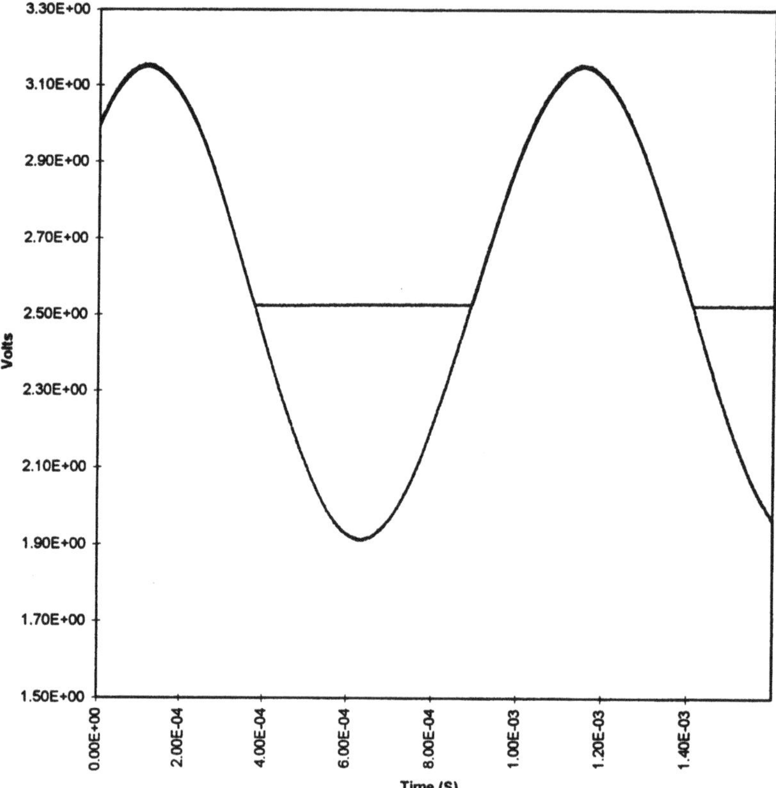

Fig. 18. Half wave rectifier input and output traces.

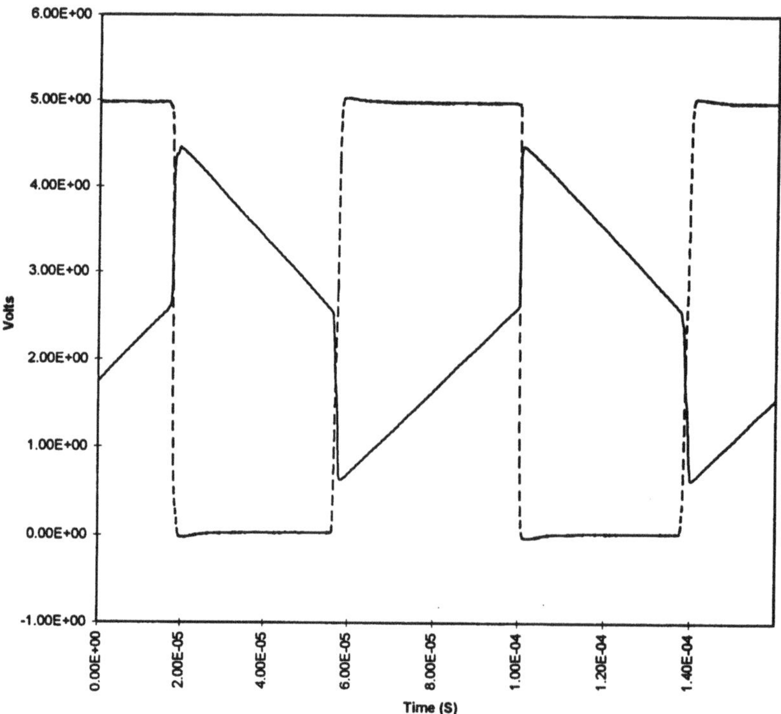

Fig. 19. Relaxation oscillator output and internal node.

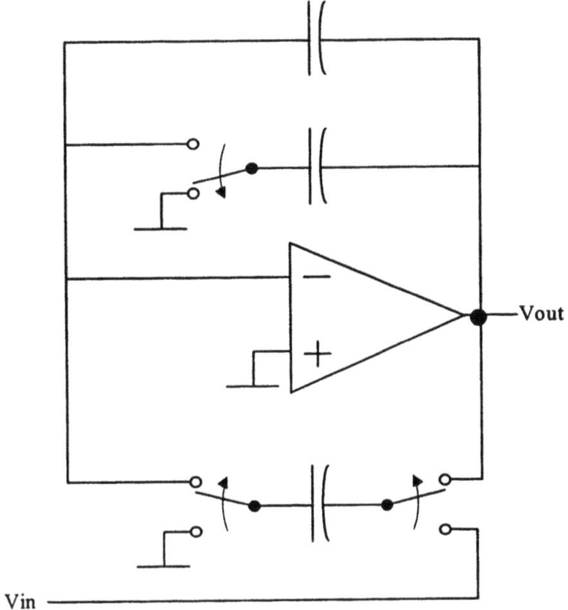

Fig. 20. Sample and hold SC circuit schematic.

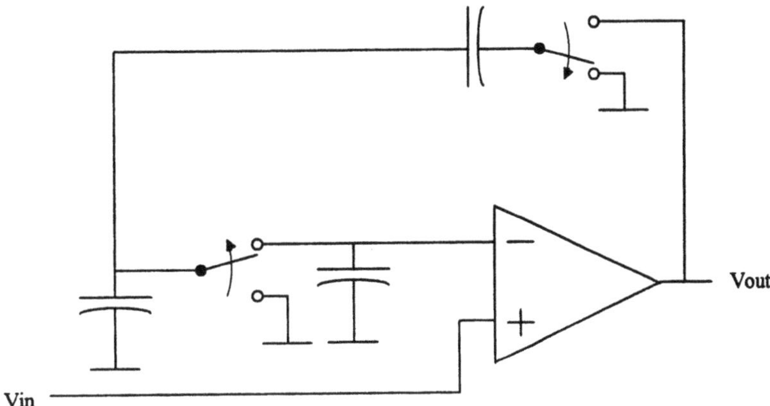

Fig. 21. Schmitt trigger SC circuit schematic.

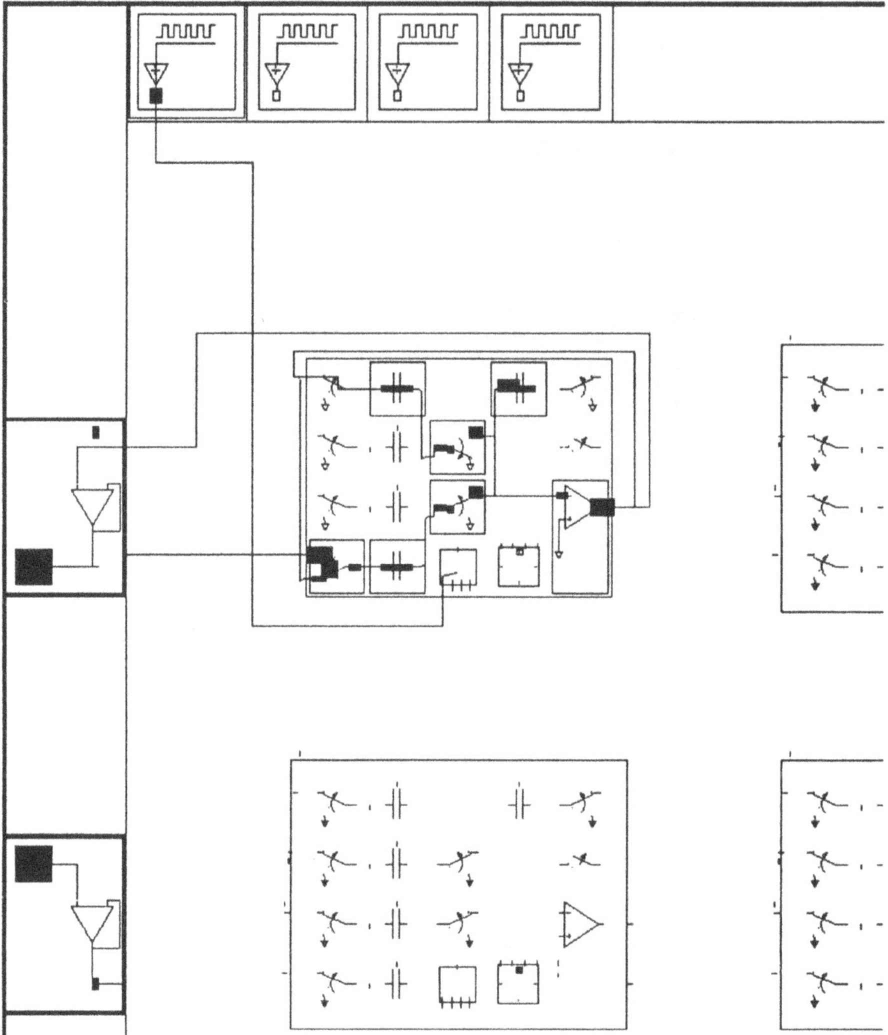

Fig. 22. Sample and hold circuit configuration.

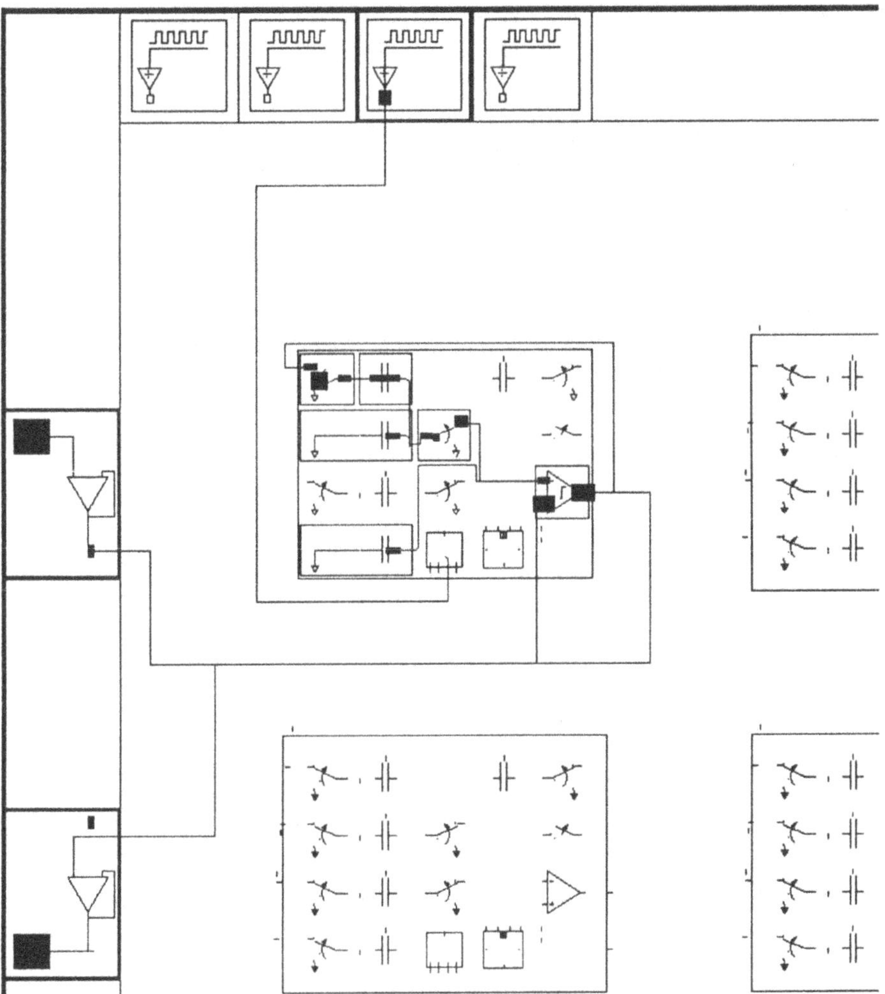

Fig. 23. Schmitt trigger circuit configuration.

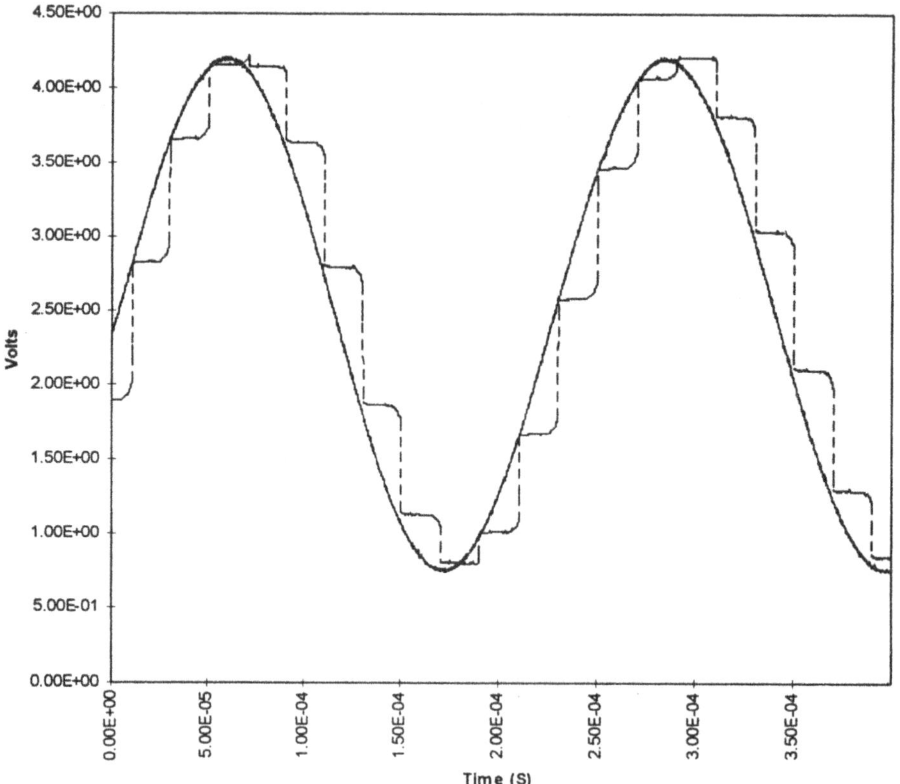

Fig. 24. Sample and hold input and output waveforms.

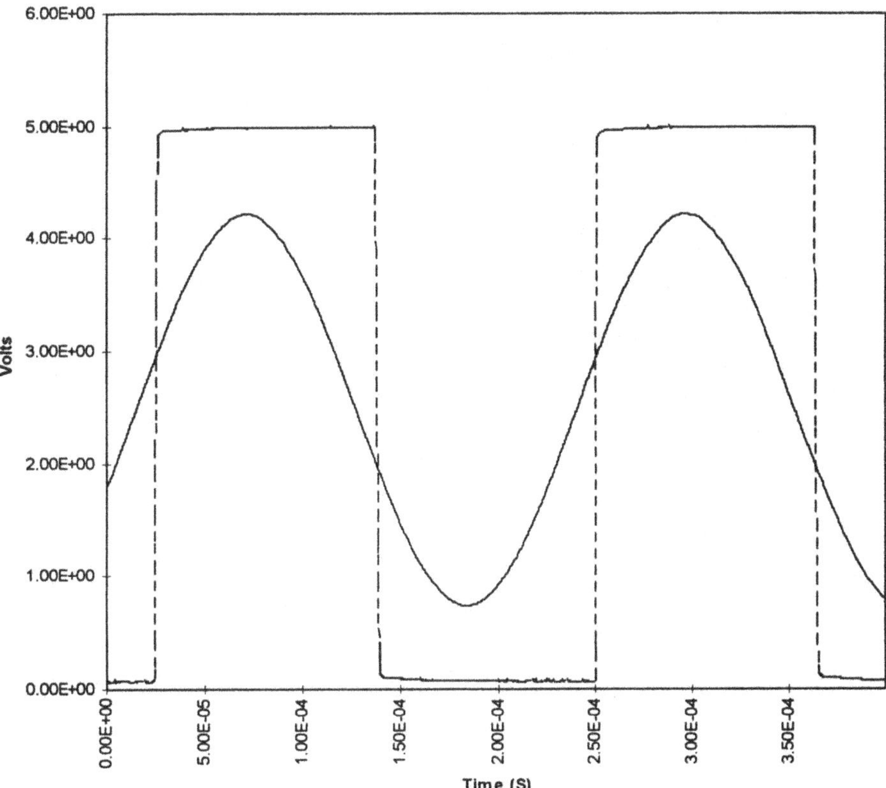

Fig. 25. Schmitt trigger input and output waveforms.

flexibility in design implementation. The DPAD2 base technology is switched capacitor because of the large number of desirable features for this application. A number of examples on the array have been demonstrated which show the basic functionality of the array in silicon.

FPAAs are now making possible the very rapid implementation of analog functions just as FPGAs did for the digital world. With field programmable analog arrays it is possible to create and implement a new design in a day or less. One of the most powerful features of such an array is that designs may be stored in software and downloaded as required. Design re-use is now possible for analog integrated circuits just as for FPGAs, with all the associated savings in time and cost.

Implementation of macro cells on the DPAD2 array allows a non-expert designer to create and implement analog functions, with little or no knowledge of the underlying switched capacitor techniques. This is a most important point because the ability to create analog solutions is now available to the non-expert. A more experienced designer is able to access the switches and capacitors directly and use the macro cells as required.

Ultimately, the goal of this work is a mixed-signal field programmable array which is created from an FPAA and an FPGA. With such a device it will be possible to create complete mixed-signal system solutions on one die with great flexibility and rapidity.

8. Appendix A

- Sample & Hold
- Unity gain buffer
- Unity gain buffer (offset compensated)
- Integrator
- Low Q Cascade Filters
- High Q Cascade Filters
- Ladder filters
- Full wave rectifier
- Relaxation oscillator
- Sine wave oscillator
- Square & Triangle wave generator
- Schmitt Trigger
- u-LAW DAC
- N-path filter
- Large time constant circuits.

Acknowledgments

The authors wish to thank Danny Bersch, David Anderson and Bill Altonen of Motorola Strategic Systems Technology, Arizona and Doug Pattullo (now of AMS) for their contribution to the DPAD2 programme.

References

1. IMP, Inc. Home Pages. http://www.impweb.com/.
2. P. Chow, P. Chowm and P. Glenn Gulak, ''A Field Programmable Mixed-Analog-Digital Array.'' FPGA'95, pp. 104–109.
3. D. Bradbury et al., ''A computational approach to VLSI analog design.'' in *Analogue Signal Processing*, Oxford Centre for Innovation, Oxford, UK, 20th November 1996.
4. J. Faura et al., ''FIPSOC: A New Concept to Mixed Signal Integration.'' in The Silicon Design Show, National Exhibition Centre, Birmingham, UK, October 1996, pp. 47–51.
5. C. Toumazou, J. B. Hughes, and N. C. Battersby, *Switched Currents—an analog technique for digital technology.* IEE: London, 1993.
6. J. E. Kardontchik, *Introduction to the Design of Transconductor-Capacitor Filters.* Kluwer Academic Publishers: Dordrecht, 1992.
7. M. Banu and Y. Tsividis, ''Fully integrated active RC filters in MOS technology.'' *IEEE Journal of Solid-State Circuits* 18(6), pp. 644–651, December 1983.
8. R. Gregorian and G. Temes, *Analog MOS integrated circuits for signal processing.* J. Wiley & Sons: New York, 1986.
9. K. R. Laker and W. M. C. Sansen, *Design of analog integrated circuits and systems.* McGraw-Hill: New York, p. 782, 1994.
10. P. E. Allen and D. R. Holberg, *CMOS Analog Circuit Design.* HRW: New York, p. V, 1987.
11. Cadence Design Systems, Inc., 555 River Oaks Parkway, San Jose, CA 95134, USA.
12. P. E. Allen and D. R. Holberg, *CMOS Analog Circuit Design*, HRW: New York, p. 204, 1987.
13. Y. Tsividis and P. Antognetti, *Design of MOS VLSI Circuits for Telecommunications.* Prentice-Hall: New Jersey, p. 334, 1985.
14. R. Gregorian and W. E. Nicholson, ''Switched-capacitor decimation and interpolation circuits.'' *IEEE Tran. Circuits Syst.* CAS-27, pp. 509–514, June 1980.
15. R. Gregorian and G. Temes, *Analog MOS integrated circuits for signal processing.* J. Wiley & Sons: New York, p. 537, 1986.
16. R. Gregorian and G. Temes, *Analog MOS integrated circuits for signal processing.* J. Wiley & Sons: New York, p. 443, 1986.
17. R. Gregorian and G. Temes, *Analog MOS integrated circuits for signal processing.* J. Wiley & Sons: New York, p. 549, 1986.
18. R. Gregorian and G. Temes, *Analog MOS integrated circuits for signal processing.* J. Wiley & Sons: New York, p. 551, 1986.
19. R. Gregorian and G. Temes, *Analog MOS integrated circuits for signal processing.* J. Wiley & Sons: New York, p. 447, 1986.
20. Dieter Seitzer et al., *Electronic analog-to-digital converters.* J. Wiley & Sons: New York, p. 139, 1983.

21. Horowitz and Hill, *The art of electronics*. Cambridge University Press: Cambridge, p. 126, 1980.

Dr. Adrian Bratt is currently working at the Motorola Programmable Technologies Centre, England as an analog IC designer. His research interests include analog signal processing using reconfigurable building blocks and analog circuit design applied to mixed-signal interfacing. He received the degree of Ph.D. from Manchester University in 1991, the degree of M.Sc. from UMIST in 1989, and a degree in Physics and Geology from Manchester University in 1988. He is a member of the IEEE and an associate member of the IEE.

Ian Macbeth received the B.Sc. degree in Electronics and Electrical Engineering from the University of Edinburgh in 1985. From 1985 to 1990 he worked at the Philips Research Laboratories in Redhill, England developing analog IC design techniques, including switched-currents. Since 1990 he has worked for Pilkington Microelectronics Ltd (now Motorola Programmable Technology Centre) developing field programmable analog array (FPAA) architectures. Currently he is the Analog Project Manager at MPTC and is a member of the IEEE.

Analog Integrated Circuits and Signal Processing, 17, 91–103 (1998)

The EPAC Architecture: An Expert Cell Approach to Field Programmable Analog Devices

HANS W. KLEIN
Lattice Semiconductor Corp., Hilloboro, OR, USA

Received July 2, 1996; Accepted June 12, 1997

Abstract. This paper describes the architectural configuration and various design trade-offs of the Electrically Programmable Analog Circuit (EPAC™), an expert-cell approach to meeting the market needs for an analog counterpart to the digital FPGA. The paper provides an overview of the technology, discusses architectural issues, and describes the internal operation of the first commercial EPAC devices. The paper concludes with various application examples and performance measurements.

Key Words: field programmable analog IC, in-system programming, high-performance cells

1. Introduction

Field-programmable logic devices, such as PLDs and FPGAs, have become a mainstream technology for the development and the production of digital functions. The biggest advantage to the end user is the integration of custom functionality at the cost structure of a standard product. The user- (or field-) programmability allows shortest implementation time of circuit ideas and immediate correction and design iteration. In combination with advanced design tools, the development of logic ICs based on FPGAs has become fast, easy, and flexible.

Another dimension of flexibility has been enabled by electrically *re*programmable devices providing the additional benefit of in-system programming. This reconfiguration of a circuit on-the-fly can make a system responsive to a certain set of conditions without implementing different logic for each of the conditions.

Key to the success of FPGA-like devices is the fact that the underlying building block is a general-purpose digital cell which, upon proper configuration, allows for the implementation of just about any digital function. In fact, that basic cell could be as simple as a 2-input NAND, even though most FPGAs use more complex building blocks. The primary concern for the user is then to have access to enough of those cells and to meet the speed requirements of the system. As a result, the digital industry has been focusing on providing programmable devices with ever growing complexity and increasing operational speed. The consequential challenge for the FPGA developers is to provide maximum logic versatility and routing efficiency.

In the analog world, the idea of a field-programmable array has been lingering in the minds of analog IC experts for many years, with the obvious objective to provide FPGA-like benefits to the users of analog functions. Even though functions such as operational amplifiers are most commonly used building blocks in analog systems, they don't implement a useful function unless components, such as resistors, capacitors, switches, or other elements are added. Thus, to achieve general-purpose functionality, a field-programmable analog array (FPAA) would have to offer a sufficient number of such components surrounding opamps or OTAs. Well known by analog IC designers, this approach introduces a large number of undesirable parasitic effects. Thus, the first challenge lies in identifying a basic building block which has enough versatility to implement a large number of analog functions.

A second key challenge lies in providing competitive building-block performance meeting the requirements of an analog market which needs products for DC to RF, micro-power to high power, low-resolution/high-speed to high-resolution/low-speed applications, and more. Because analog performance characteristics outnumber those of digital cells 10:1 and more, it is substantially more difficult to provide analog functions which indeed meet the *all* the users' requirements.

A simple example is that of a two-transistor inverter whose primary digital characteristics are propagation delay and average power consumption. The *same* circuit used as a simple analog amplifier must meet specifications such as gain, offset, linearity, PSRR, noise, slew rate, power consumption, bandwidth, distortion, stability, and drift. Many of these parameters can be critical and thus must be either programmable or at least well controlled under all programmable circumstances. Obviously, many of these parameters are correlated which means no single building block can meet all these vastly varying requirements.

It is these two basic issues (i.e., building-block versatility and range of performance characteristics) that create the classical FPAA dilemma: broad functionality at limited performance or reduced functionality at competitive performance.

Even though the first option promises that a user can implement any function needed, this solution is likely to never make it into production unless all of the performance requirements are acceptably met. This renders this approach economically non-viable.

On the contrary, special-purpose products with a high degree of programmability have already been introduced to the market as these products can indeed offer sufficient performance given their focused functionality. One of the first commercial examples for this type of product is an electrically programmable switched-capacitor filter product introduced by Sierra Semiconductor in 1989 [7]. Another example of electrically programmable products are A/D converters with variable resolution. All of these products, though allowing to change certain characteristics, implemented a single fixed analog function.

Researchers have continued to search for solutions for viable FPAA circuit techniques and building blocks. Finally, in early 1996, the first FPAA-type product became commercially available in the form of EPAC (Electrically Programmable Analog Circuit) devices. Unlike its predecessors, this device is programmable in performance characteristics, interconnect, and functionality. In the remainder of this paper, we will describe the technological and architectural difference that make this product series commercially viable.

2. EPAC Overview

EPAC devices are analog ICs that can be programmed in functionality, interconnect, and performance characteristics. The EPAC device architecture is based on a mixed-signal CMOS process with on-chip configuration shift register (CR) and EEPROM memory to provide both permanent configuration and in-system reprogrammability.

An important aspect of the EPAC approach is that it raises the actual design task from a tedious and error-prone component level to a functional, or block, level (see Fig. 1). These modules are key to the usage of analog functions as the majority of users don't have the skills, time, or experience to successfully implement integrated analog functions at a component level.

As opposed to providing user access to all low-level components to form certain functions, an "expert cell" approach has been chosen wherein the building blocks feature a predefined set of high-level functions with programmable performance characteristics. This corresponds to a "macro" approach in digital FPGAs which incidentally is now becoming more popular as higher-level functions, such as fast multipliers and memory modules, need to be integral parts of an FPGA architecture. "Expert cells," in our context, are modules which implement a certain set of programmable functions in a controlled environment. As a consequence, the expert cell can be designed to work under all programmable circumstances permitted within the cell boundary even considering the significant number of parasitic effects that must be dealt with in programmable analog circuits.

Even though each expert cell already provides programmable characteristics and functionality, additional (i.e., system-level) functionality is then provided by programming the interconnect between individual expert cells. One major issue is to design these expert cells such that their individual performance is not affected by system-level interconnect choices made by the end user at a later time. This essentially corresponds to electrically decoupling the building blocks from each other. As a result, each expert cell can now be developed independently and eventually becoming part of a larger FPAA-like system.

An example of an expert cell would be a programmable D/A converter (DAC) where the speed, power-consumption, reference input, and hence

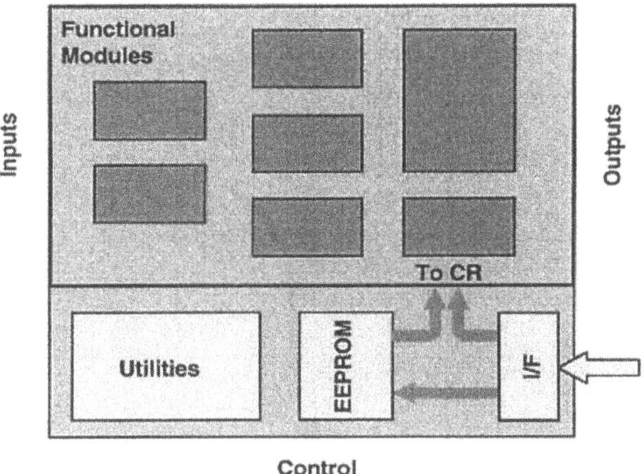

Fig. 1. General EPAC framework.

output range are user programmable. The DAC expert module is then designed to perform as specified under all circumstances allowed by the system architecture. An example for various system-level configurations would be using the DAC for offset correction in an input stage, for providing a stable reference for a sensor biasing module, or a threshold for a comparator.

Using a PC-based development system (''Analog Magic'' in our case), the user defines a system by selecting the functionality of the modules required along with their respective performance settings and module interconnections. Upon downloading the resulting configuration file into the EPAC device, the function as seen on the computer screen is implemented in silicon in a matter of seconds. This then finally provides the desired benefit of simple and fast design implementation of user-programmable analog functions.

Selected elements of EPAC technology, i.e., memory, architectural framework, expert-cells, and interconnect, are now discussed in greater detail.

3. The EPAC Device Framework

EPAC devices are built on a framework which incorporates the functional modules together with programming features, debugging aids and interconnect ''highways.'' The key elements of that framework (see Fig. 1) include a serial interface, an EEPROM-memory module, a ''utility'' section, and a functional section containing the programmable modules.

An expanded view of the interface circuit is provided in Fig. 2. The serial interface allows the user to both configure the IC and to send commands, including probing of internal analog or digital signals, readback of configuration data, control of power management, and other functions. The interface also handles communication with other EPAC devices in a cascade, or daisy-chain, arrangement which can then be programmed in sequence or individually via the same serial interface wires if a microprocessor is available in the system. In such case, the framework also performs automatic synchronization among multiple EPAC devices so that all ICs ''wake up'' in the new configuration simultaneously.

Embedded in the EPAC framework, the pool of analog modules can be accessed and configured using the on-chip memory. The framework provides two independent memory sections, i.e., an EEPROM module which acts as ''shadow memory'' and the shift-register type configuration register (CR). These will be described in the following section.

4. Memory Modules and Process Selection

All initial EPAC devices are based on a 1.2 micron analog EECMOS process. This required adding a self-contained EEPROM process module (including a high-voltage generator) to a proven analog CMOS process without affecting critical analog performance

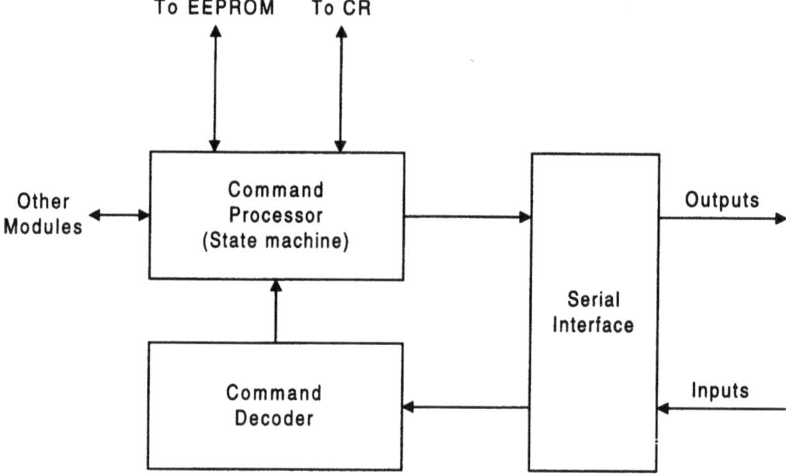

Fig. 2. Interface architecture.

characteristics. This permits the expert-cell design engineers to draw from a library of proven analog models and high-performance circuitry, ranging from very fast circuits to high-precision converters, references, etc. This also ensures that the performance of the modules embedded in EPAC devices are competitive with that of dedicated ICs, while offering higher degrees of flexibility and functionality.

The EEPROM cells were designed for robustness, not density. Thus, a differential cell structure (see Fig. 3) was chosen allowing for maximum process variation, large noise and temperature margin, and long data retention. The memory module is organized in an 8-bit word format with page erase. When writing data to the memory, an automatic page erase is issued followed by writing all configuration bytes, not just individual ones the user might want to change.

In addition to the EEPROM, a segmented static configuration register (CR) holds the configuration data that actually controls the function and characteristics of each module. The command processor has access to selected segments within the CR for fast access to important subsections, such as an offset register. Upon power-up, the command processor causes the EEPROM to download its contents into the CR, then wake up the chip. Thus, the EEPROM acts like a backup memory and, in fact, the user can selectively change just the contents of the EEPROM, or that of the CR. Writing to the EEPROM requires 10 ms per byte while writing to the CR takes place in 8 μs per byte. A chip like the IMP50E10 can thus be completely reconfigured in 200 μs.

5. "Expert-Cell" Modules

Unlike in FPGAs, where generic logic modules can implement virtually any digital function, analog

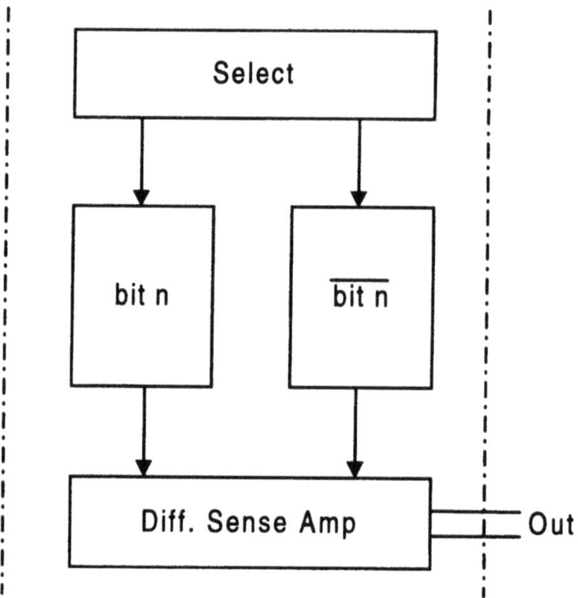

Fig. 3. Differential bit-slice (EEPROM).

circuits must be dedicated to address the large variety of specifications, such as input and output voltage and current, frequency range, noise and accuracy requirements. Since each of these specifications require their specific trade-offs (area, cost, power-consumption, etc.) it is more challenging to define a programmable analog architecture than a digital FPGA. Even though it is possible to create a rather generic user-programmable array architecture [1–3], the wide range of performance requirements makes it necessary to optimize the ''array'' for a certain class of applications. As a result, EPAC devices contain modules that are optimized for a certain class of applications and hence offer better performance-functionality trade-offs than general-purpose arrays. The same optimization also simplifies the chip architecture and minimizes circuit overhead and, hence, cost. Examples of the kind of application areas for this approach include general signal conditioning, data communication, data/signal monitoring, process control, etc. Consequently, EPAC devices contain modules that are optimized for use in those application classes.

One of the key differences between EPAC devices and general-purpose FPAAs lies in the granularity of the building blocks, as indicated in Fig. 4. While FPGAs and FPAAs provide access to an array of relatively low-level cells, EPAC devices provide access to high level cells (or ''macro'' modules) generally arranged in a non-array fashion. The cells typically include high-level functions such as D/A converters, amplifiers and multiplexers rather than individual operational amplifiers, resistors, capacitors, switches, or other such low-level components.

Each of these high-level modules have been given a number of programmable functions and characteristics all of which are guaranteed to work, no matter how they are configured or how they are connected to other macro cells. Hence, the user (or, for that matter, the CAD software) does not need to worry about temperature matching, offset cancellation, parasitic coupling, stability, loading effects, gain-bandwidth trade-offs, minimization of power consumption, etc. By contrast, analog arrays following the FPGA paradigm suffer from parasitics introduced by often unpredictable routing and signal coupling, and the cells cannot be optimized for specific performance requirements (such as low noise) while, on the other hand, they offer the potential for a larger number of implementable functions.

Fig. 5 shows, for overview purposes, a simplified block diagram of an EPAC device, the IMP50E10, which is targeted at general signal-conditioning applications. The ''flow'' of the chip is from input modules on the left, through core modules to output modules on the right. The shaded area at the bottom of the diagram represents the ''utility section'' mentioned earlier which contains the serial interface, memory module, power management, oscillator, probe, and an operational amplifier. The routing channel has also been indicated in a simplified fashion to maintain overview. Each module can be connected to any other module, or several of them.

The output module of the IMP50E10 (three are available on this chip) is a good example to show the degree of flexibility which can be designed into an EPAC expert cell module: It can act as an amplifier with reference to ground (e.g., 0 V), or virtual ground (e.g., 2.5 V), with or without a low-pass filter in the signal path. The amplifier can be ''free running'' or a sample-and-hold type. The module can also act as a comparator with and without hysteresis. A second

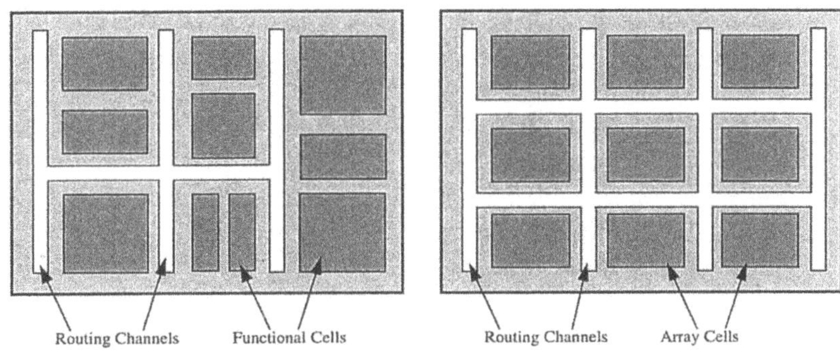

Fig. 4. Comparison of EPAC topology with that of an analog array.

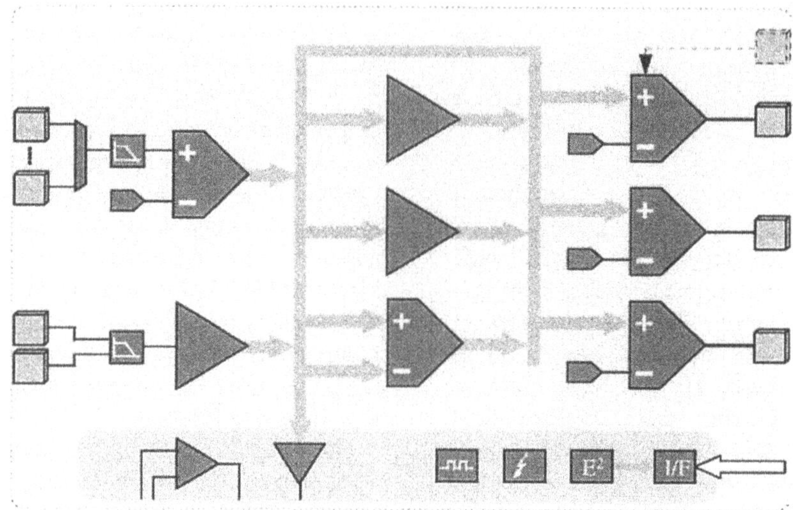

Fig. 5. Simplified block diagram of the IMP50E10.

input to the module connects to a dedicated 5-bit D/A converter, letting the user define the trip-point of the comparator. The same DAC input can be used to either drive the amplifier (thus making it a reference module) or provide for DC-shift of the outgoing signal. This output module can operate at a reduced power consumption for slower-bandwidth signals. The module can also be shut off altogether, either selectively while other modules continue to operate, or globally when the entire chip powers down. With all of these combinations, the user doesn't need to worry about component-level detail as this has been already implemented.

6. Signal-Highway Interconnect Approach

One of the unique features of EPAC devices is the user-programmable interconnect among various analog modules. This allows the user to implement different functions with the same chip. Likewise, some modules may be reconfigured automatically without the user even realizing it. An example for the latter would be the automatic reconfiguration of several modules inside the IMP50E30, a monitoring and diagnostics device. There, a programmable-gain amplifier, a DAC, a comparator module, and a reference module are temporarily reconfigured on-the-fly, transparent to the user, to switch from a monitoring configuration to an A/D converter configuration. This reconfiguration can be limited to the settings of a certain module, or include the rerouting of signals.

Programmable interconnect is implemented by an on-chip "signal highway" (see Fig. 6) to which all functional modules have access. This "signal highway" is an on-chip bus of differential, balanced, and shielded signal wires. Attention must be paid to optimizing this signal bus as signals of various amplitudes and frequency content are likely to coexist on the same bus. For example, if a signal is routed with all amplifiers in series, the total gain on the IMP50E10 reaches 20,000 which means signals on neighbor-tracks have vastly different amplitude and phase. Non-segmented, minimum-capacitance and fully balanced bus lines are the best approach to minimizing undesirable signal crosstalk. Note that the underlying EPAC framework is not a regular grid-array of a large number of general-purpose cells (like in FPGAs) but rather an assembly of optimized modules. Hence, it is possible to predict all possible and meaningful interconnect schemes. Consequently, the signal highway can be optimized for 100% routability with lowest crosstalk and minimization of other parasitic effects. This is an important difference to the true array approach because the traditional problem of severely limiting the performance by the programmable interconnect network is not present in EPAC devices.

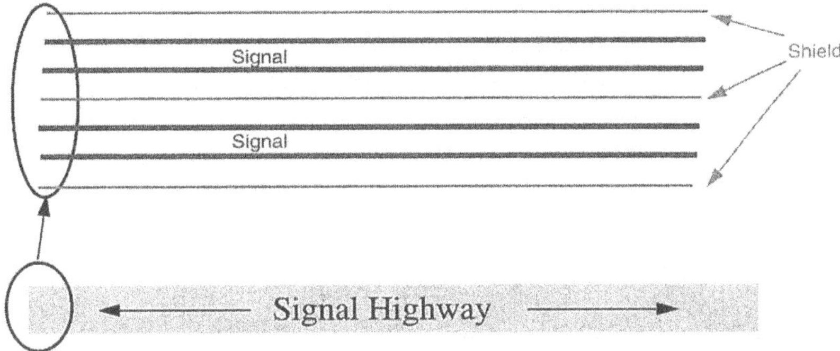

Fig. 6. Signal highway implementation detail.

Fig. 7 shows how functional modules can tap into the signal bus under control of the configuration registers. These analog switches are arranged in clusters to form "highway-connectors" at the appropriate locations. Inside each functional module, analog switches also select component values or ratios, feedback tap-points, change bias conditions, and perform all other analog programming. All of these configuration-related control signals are provided by a set of static control bits stored in the respective portion of the configuration register (CR). The output signals of that register are input to a configuration management unit embedded in each functional module. The benefit of this approach is that only a small number of control bits must be made available to each functional module which also helps to minimize routing overhead and disturbance of the signal highway.

7. Intermodule Communication

As indicated in Fig. 7, some modules can communicate with other modules. This is most helpful when the reconfiguration of several modules is involved to execute a more complex function, such as an auto-zero offset calibration. The intermodule communication feature allows the command processor to control just one status bit or byte which in turn causes all modules involved to take the state necessary for that function. To illustrate, an offset-calibration command issued by the user will set one status register triggering the following events: redirect the input multiplexer to disconnect the input signal and short the selected inputs, redirect the selected output module to switch to linear amplifier mode, disable the smoothing filter if it was set (for faster response), and apply a certain reference voltage to the shifter

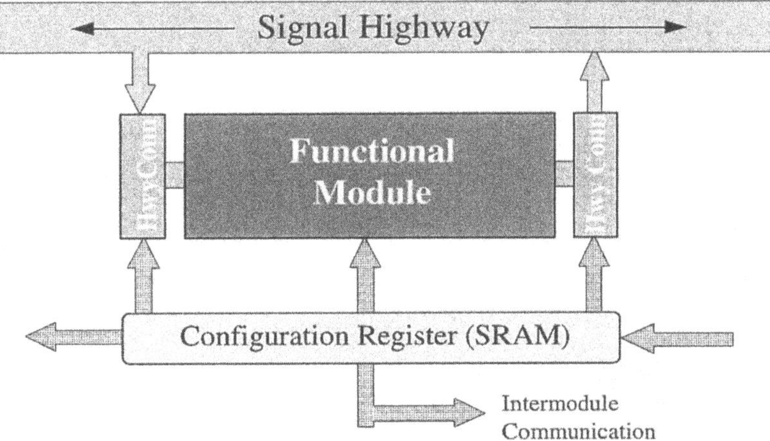

Fig. 7. Functional module and its surrounding support modules.

stage. Then the auto-zero state-machine starts searching for the DAC value needed to compensate for the present system offset. Upon completion, all modules switch back to their previous state and resume normal operation, with all offset removed.

8. Analog Module Implementation

Because of the target applications of general-purpose signal conditioning switched-capacitor (SC) technology was chosen as the prime implementation technique for the IMP50E10. SC modules offer advantages of high reproducibility and excellent stability, and these techniques have been in use for many years. Fig. 8 shows the underlying principle of SC circuit modules.

Because this technique is well understood in the analog IC design community, its basic operation is not discussed here. It is noted, however, that the circuit implementation of a real ''expert cell'' is far more complex than shown here. For example, to increase flexibility and performance all capacitors can be made programmable, multi-phase clocking can enable offset and noise cancellation through ''correlated double-sampling,'' clock rates can be altered to allow for various speed settings, and multiple inputs can be used to sum multiple inputs. In addition, modules are best implemented in fully-differential manner, effectively doubling the feedback circuitry with the substantial benefit of much improved power-supply rejection and excellent clock-feedthrough cancellation. The same differential technique also allows for simple inversion of signal polarities which is essential for subtracting signals or inverting transducer-signal polarities.

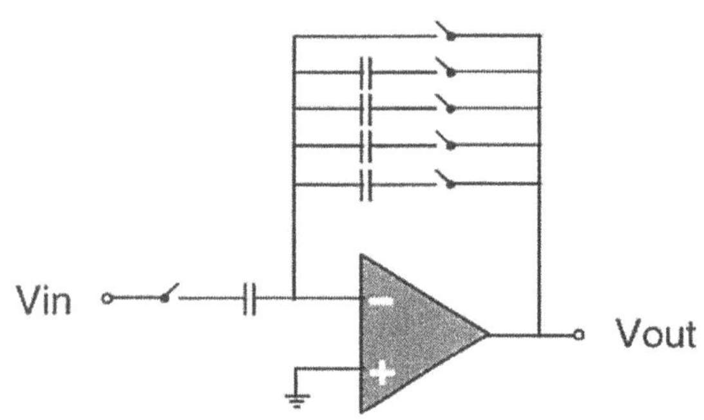

Fig. 8. Switched-capacitor implementation concept.

A more detailed schematic of an SC amplifier module is given in Fig. 9, showing differential architecture and programmable input capacitors C1a and C1b. The gain of this circuit is simply C1/C2. Unlike in SC filters where integrators are used where charge is accumulated, all capacitors get reset to an initial value in each clock cycle. As typical in many SC circuits, the output signal is valid only in one phase.

One important benefit of SC technique is that it is relatively easy to operate off of a 5 V supply while input signals can go from rail to rail, and even beyond, without compromising performance. Though continuous-time circuits offer higher bandwidth, the need for programmable gain, differential inputs, and reasonably high input impedance require significantly more circuitry than in SC technique. On the downside, because SC circuits are clocked, the Nyquist sampling theorem must be considered. Thus, the bandwidth of the incoming signal must be limited.

Important design considerations for an SC amplifier include excellent matching between input and feedback capacitors for high gain accuracy while high common-mode rejection requires excellent matching between inputs capacitor pair and the feedback capacitor pair. In other words, all capacitors have to matched to each other extremely well to meet both gain-error and CMRR requirements. In reality, this requirement is difficult to meet as layout constraints generally force an array structure where C2a is embedded within C1a, and likewise C2b is embedded within C1b. This means that the upper and lower half of the differential structure are not as well balanced as

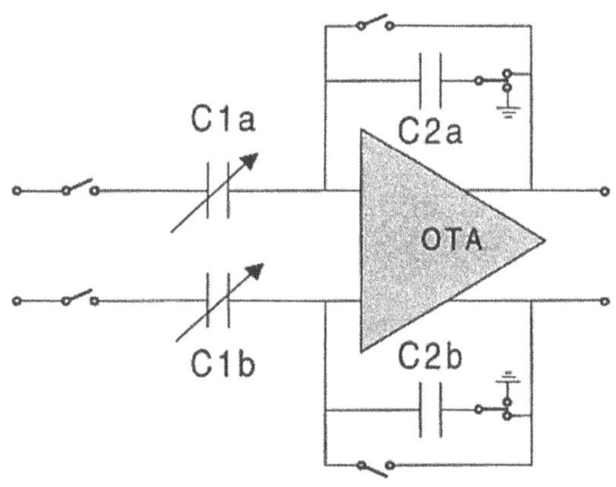

Fig. 9. Amplifier module.

desirable. In practice, matching between the differential branches around 0.2% has been achieved. This limits the CMRR to about 54 dB which means that applications requiring common-mode higher rejection need a different circuit architecture or, at least, a different layout. Thus, the circuit as shown in Fig. 9 is best used as a ''core'' module where common-mode variations are very small.

Fig. 10 shows the general architecture of an IMP50E10 output module. This is an example of a multi-functional module using both SC and continuous-time circuitry. Since the internal signals coming from SC modules are clocked while most users prefer off-chip signals to be continuous-time, a Track&Hold (T&H) amplifier picks up a differential signal from the signal highway thus providing a continuous-time output signal. Still differential, that signal drives a programmable amplifier/comparator section comprising an input stage, and output driver, and a feedback network. The final output signal is single-ended, driven within milliVolts of the rails, short-circuit protected, and unconditionally stable. The feedback network determines the gain of this differential-to-single converter and the characteristic of a low-pass filter circuit for smoothing the signal samples that exhibit the staircase shape typical to SC circuits. When in comparator mode, optional hysteresis can be activated. Not shown in Fig. 10 for clarity, there are additional inputs for shifting the analog signal, for setting the threshold for the comparator, and for the control of the driver's slew rate.

9. Application Examples

To illustrate the usage of the IMP50E10, Fig. 11 shows an implementation of a multi-sensor analog front end for an industrial battery-charger application. This implementation supports in-system reprogramming. One micro-controller can serve a number of these analog front ends, because of the minimal interaction required with any one front-end circuit.

In the example shown, the IMP50E10 picks up different input signals (voltage, current, temperature) and conditions them for A/D conversion. The ''zoom-and-pan'' function provided through offset-shifting and amplifying within the IMP50E10, allows the ADC to cover a much larger dynamic range than otherwise possible. Thus, an embedded 8-bit ADC, as used in this example, can easily span a 16-bit dynamic range with the same resolution as a 16-bit ADC.

The temperature signal is derived from a thermocouple although a different (and cheaper) sensor could have been used here. Unlike the voltage and current signals which are selected through the input multiplexer, the temperature signal enters the chip in a

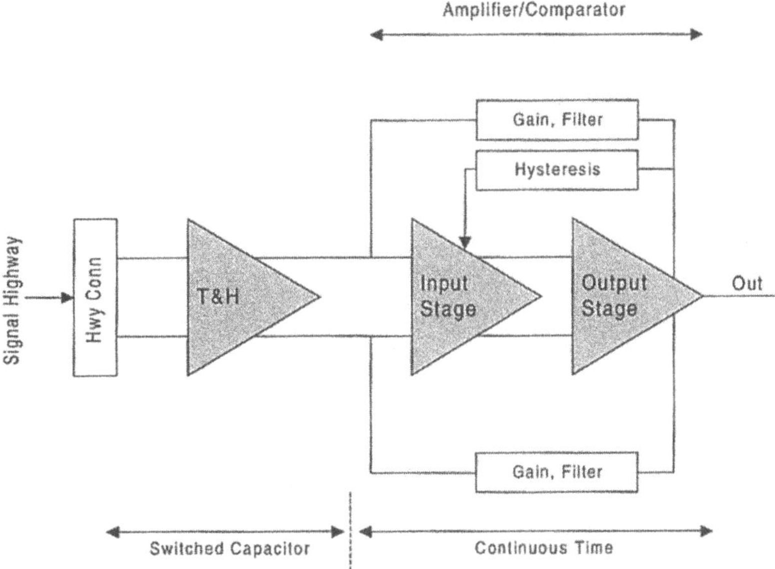

Fig. 10. Output module architecture (IMP50E10).

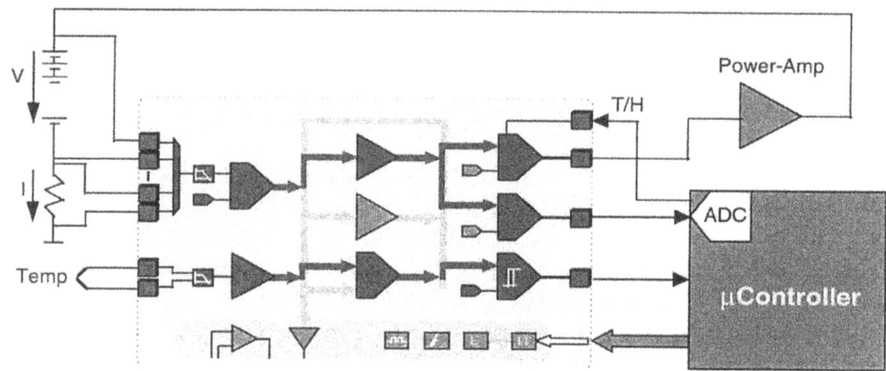

Fig. 11. Application example using the IMP50E10.

separate, non-multiplexed, channel. Also, unlike the other signals, the temperature signals triggers an on-chip comparator at a programmable trip point, only alerting the microcontroller when the temperature exceeds a critical limit.

Other system-level functions not obvious from Fig. 11 include automatic offset cancellation for chip and sensor offsets, as well as automatic changes of gain, offset, and filter settings when switching among the different input signals.

Another application example shows the IMP50E30 in a system-monitoring application (Fig. 12), such as for server computers and high-end PCs. This application presents the challenge of providing both small and large signals, even exceeding the device's 5 V supply. Optimized for working such environment, the IMP50E30 features on-chip high-voltage attenuators/shifters at some of its inputs allowing inputs of up to $+/-17$ V without external components, while it runs off of a 5 V supply. These special input structures are a good example why it is so difficult to build a truly ''general-purpose'' analog array satisfying such vastly varying needs of the analog market. The device also features a programmable logic (state-machine) module which allows automatic scanning of 16 channels with individual gain and filter characteristics applied to each pair of signals. Those conditioned signals are then subjected to a window-comparator applying programmable thresholds to each of these signals. This type of on-the-fly reconfiguration capability is inherently efficient with EPAC devices considering that for each new set of configuration data only the an incremental amount of memory must be available.

Even though the device normally operates fully independently in the ''background,'' the host CPU

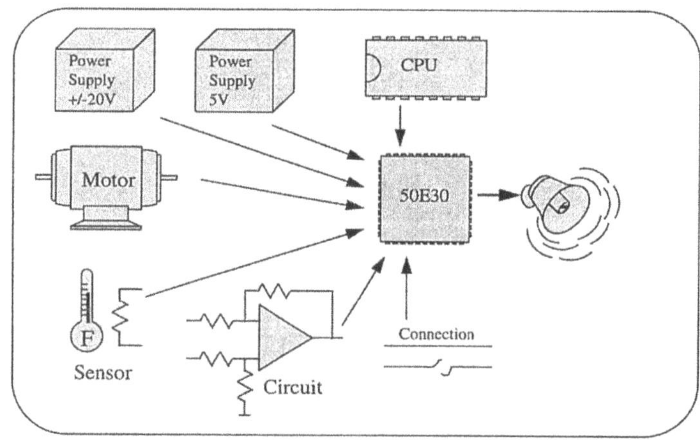

Fig. 12. Monitoring application with the IMP50E30.

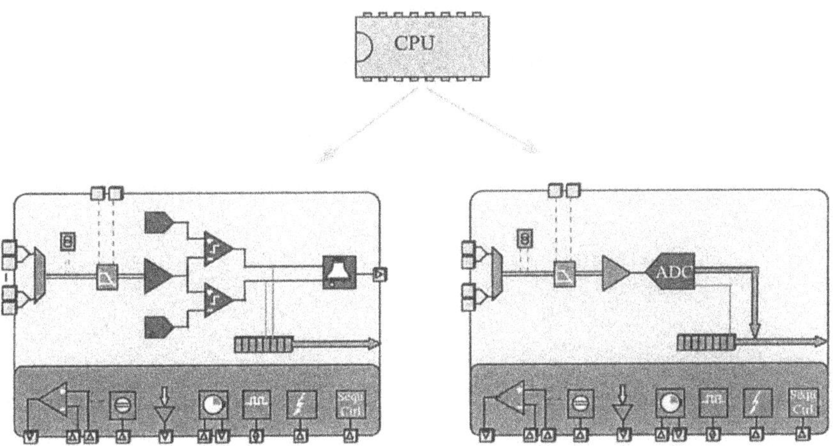

Fig. 13. Automatic reconfiguration of multiple modules.

will be alerted in case of an ''out-of-range'' condition. If desired, the CPU can then take appropriate action, such as alerting the system administrator of an impending problem. In one mode, the CPU can simply reconfigure the device into a diagnostics mode with a single 8-bit command via the serial interface, as described earlier. The new configuration then allows for A/D conversion of any signal thus providing more detailed information for further decision by the computer or the user.

The modules involved in the automatic reconfiguration are the programmable instrumentation amplifier, the comparator(s), the DAC(s), the sequencer and other logic blocks as indicated in Fig. 13. This is also an example of automatic reconfiguration which is controlled by the device itself, without user intervention. To simplify the control of all these reconfiguration events, module-to-module communication is employed, as mentioned before.

10. Performance Summary

The IMP50E10 and IMP50E30 have been implemented in switched-capacitor technology which requires a high-quality clocking scheme to minimize undesired clock-feedthrough, signal-recovery glitches, and offset errors. The devices have an on-chip multi-phase oscillator with controlled slew-rate which drives all SC module simultaneously, without refresh. This guarantees highest clocking quality. At normal speed settings, the clock module sets the maximum sampling rate to 250 kHz. The oscilloscope photo in Fig. 14 gives an indication of the signal quality at one of the IMP50E10 output modules without the use of smoothing filter. The upper trace is the input signal (at 1 kHz), the lower trace is the amplifier (400 ×), unfiltered output signal swinging within 200 mV of the rails. One can see that the circuit techniques employed greatly suppress the ''switching noise'' commonly found in SC circuits.

The second scope photo (Fig. 15) is also taken from an IMP50E10 application. Here, the input signal is a 10 kHz sine wave (top trace), the analog output signal after low-pass filtering in both the input and the output (smoothing) modules is shown in the center. The resulting phase delay is easily recognized. The digital output signal (bottom trace) has been derived from running the input-filtered and amplified sine wave

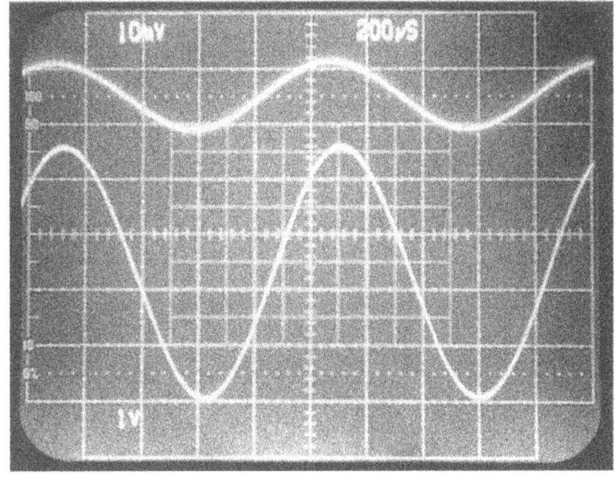

Fig. 14. Input and output signals (IMP50E10).

Table 1. IMP50E10 performance summary.

Power Supply Voltage	5 V +/− 10%
Quiescent supply current	3.8–18 mA (configuration dependent)
Sleep-mode current	40 μA
Max. sampling rate	250 kHz
Max. signal bandwidth	125 kHz (Nyquist rate)
Input/output voltage range	0–5 V (rail-to-rail)
Gain range and error	1–20,000 V/V, 1.2% typ.
Gain drift	30 ppm/°C
Offset before/after Auto-Zero	< 10 mV, < 100 μV
Offset drift	50 μV/°C
Max. capacitive load	infinite (unconditionally stable)
Noise level, DC-15 kHz	0.4 μV/sqrt(Hz), input referred
Linearity	> 12 bit @ 0.5–4.5 V

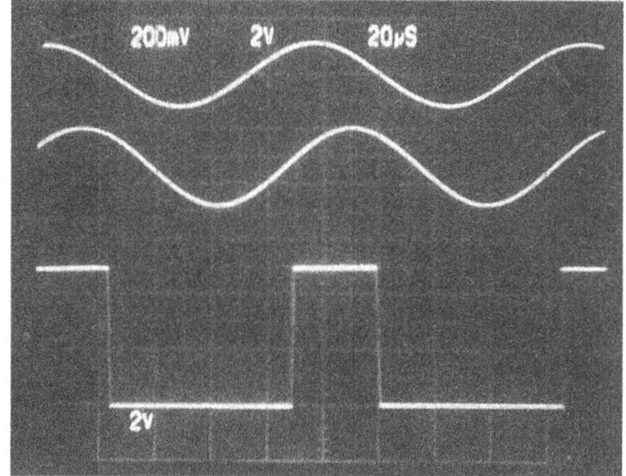

Fig. 15. Input and output signals (IMP50E10).

through a second output module configured as comparator with a threshold set to 3.5 V.

Measurements have been performed to verify that the linearity required by 12-bit systems can be met by the IMP50E10 modules. The graph in Fig. 16 shows the linearity error from 0.5 V to 4.5 V at 5 V supply for five different devices. An error of 0.5 mV corresponds to 1/2 LSB in a 12-bit system. The configuration involved one input and one output module, the output filter was activated and the system gain was 20 V/V. The roughness of the graphs was caused by noise present in the test system.

Table 1 summarizes some of the key performance characteristics of the IMP50E10.

11. Summary

In this paper we presented the expert-cell approach to programmable analog solutions for rapid prototyping, development, and customization of analog integrated circuits. In contrast to general-purpose grid-array cells, the expert-cell approach allows for higher performance, less overhead, and lower cost.

As a result of operating in the analog domain, EPAC devices are more tailored to a certain range of applications than digital FPGAs, but they basically offer the same benefits. The high degree of programmability and ease of development, using

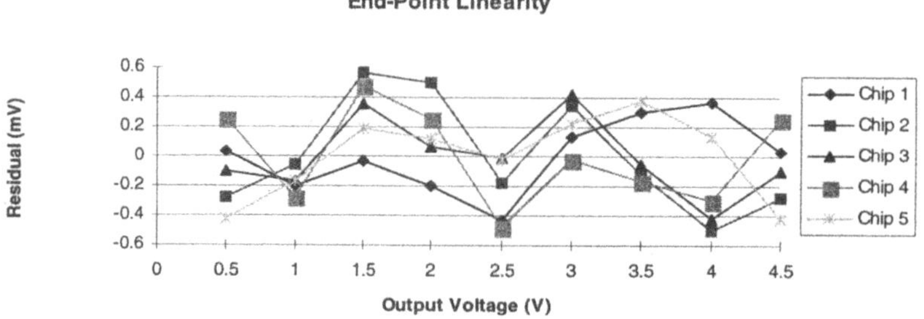

Fig. 16. Linearity error of 5 devicies (IMP50E10).

low-cost PC-based tools, allow the user to implement a complex analog function in minutes. Furthermore, in-system reconfigurability enables modifications on a system level with no, or minimal, hardware changes. This not only makes last-minute changes easy, but also enables functional reconfigurations such as self-test, system calibration, and various operational modes.

Acknowledgment

Many engineers from various disciplines have instrumentally contributed to the development of this product line, too many to be named individually. They all deserve credit for solving difficult circuit, process, simulation, software, test, and application problems. As a team they accomplished what individual contributors would not be able to do. They can all take pride in their achievements.

References

1. E. Pierzchała et al., "Current-Mode Amplifier/Integrator for a Field-Programmable Analog Array." *ISSCC Digest of Technical Papers*, San Francisco, 1994.
2. E. Lee et al., "A Transconductor-based Field-Programmable Analog Array." *ISSCC Digest of Technical Papers*, San Francisco, 1994.
3. Pilkington, Conf. Proceedings, PLD, London, 1995.
4. R. W. Brodersen et al., "MOS Switched-Capacitor Filters." *Proc. IEEE* 67(1), Jan. 1979.
5. R. Gregorian et al., "Switched-Capacitor Circuit Design." *Proc. IEEE* 71(8), Aug. 1983.
6. E. Habekotte et al., "State of the Art in Analog CMOS Circuit Design." *Proc. IEEE* 75(6), June 1987.
7. F. Goodenough, "Create Switched-Cap Filter IC with EEPROM/" *Electronic Design*, July 1989.
8. "Switched-capacitor analog circuits with low input capacitance." US Patent 5,617,093.
9. H. W. Klein, "The EPAC Architecture: An Expert Cell Approach to Field Programmable Analog Devices." *FPGA-Symposium*, Monterey, 1996.
10. A. Wolff, "Analog Controller for Overworked Micros." *Electronic Design Magazine*, June 24, 1996.

Hans W. Klein joined the founding team of the Institute for Microelectronics Stuttgart (IMS), West Germany, in 1983 with focus on mixed-signal circuit R&D and training of industrial engineers and managers.

In 1988 Dr. Klein joined International Microelectronic Products (now IMP, Inc.), San Jose, CA, USA, to reengineer IMP's Mixed-Signal Design Methodology and to develop ICs for mass-storage systems including the industry's first MR preamplifier in CMOS technology.

Under his leadership, IMP introduced the EPAC family of user/field-programmable analog circuits. These electrically programmable analog circuits are the first devices to be fully configurable in function, interconnect, and performance. Dr. Klein is now with Lattice Semiconductor Corp., Hillsboro, OR, USA where he is in charge of Mixed-Signal products.

Analog Integrated Circuits and Signal Processing, 17, 105–124 (1998)

A Current Conveyor based Field Programmable Analog Array

CHRISTOPHE PREMONT

CIMIRLY INSA Lyon, Bat 401, 3eme etage, 20 Av. Albert Einstein, 69621 Villeurbanne Cedex, France
premont@cegely.insa-lyon.fr

RICHARD GRISEL

CPE, LISA EP-CNRS 0092, 43 Bd du 11 novembre 1918, BP 2077, 69616 Villeurbanne Cedex, France
grisel@cpe.fr

NACER ABOUCHI

CPE, LISA EP-CNRS 0092, 43 Bd du 11 novembre 1918, BP 2077, 69616 Villeurbanne Cedex, France
abouchi@cpe.fr

JEAN-PIERRE CHANTE

CEGELY INSA Lyon, Bat 401, 3eme etage, 20 Av. Albert Einstein, 69621 Villeurbanne Cedex, France
chante@cegely.insa-lyon.fr

Received July 11, 1996; Accepted June 9, 1997

Abstract. An approach for designing a Field Programmable Analog Array (FPAA) is described. The analog array is based on current conveyors and benefits from two major interests: a large bandwidth and a low number of discrete components needed for the implementation of analog functions. An Analog Elementary Cell (AEC), based on current conveyors has been developed, and it is associated with programmable resistors and capacitors. Analog functions can be performed programming several AECs as current-mode amplifiers, analog multipliers, etc. The main purpose of this paper is to introduce current conveyor based analog blocks which are very-well suited for the implementation of FPAA. A particular interconnection architecture is addressed using current conveyors as switches. The major key feature of the proposed approach is that current conveyors are used as active elements and switching elements. A new topology based on the developed AEC is proposed and should be shortly validated.

Key Words: Field Programmable Analog Array, current conveyor

1. Introduction

A FPAA consists of a number of reconfigurable analog elementary cells which can be interconnected by a programmable architecture. If one considers the great improvment that FPGAs (Field Programmable Gate Arrays) represents for digital design, one can imagine that an analog counterpart for this type of circuit would provide electronic designer with a very efficient and powerful tool. However, FPAAs have not been developed as much as FPGAs because of a lack of real market and because a number of limitations such as precision, offsets, noise, etc.

The key feature of FPAAs are flexibility and programmability. This paper deals with the implementation of basic analog functions, investigating new programmability features and topology. Based on current conveyors, this approach provides designers with programmable elementary cells. This paper focuses more on the elementary cell of the analog array than on the analog array itself because the work is still in progress.

Firstly, Section 2 introduces the current conveyors and their applications. Section 3 describes the analog elementary cell (AEC) and the programmable elements. Then, Section 4 focuses on AEC based application in order to show the real potentialities of the proposed approach. The analog array topology and programmable features are addressed in Section 4.

2. Current Conveyor

A current conveyor (CC) [1–3] is a three terminal device which performs many useful analog signal processing functions (see Table 1) when arranged with other electronic elements in specific circuit configurations. The operation of this device is such that if a voltage is applied to input terminal Y, an equal potential will appear on the input terminal X. In a similar fashion, an input current forced into terminal X will result in an equal current flowing into terminal Z. Fig. 1 shows the schematic circuit of the second generation current conveyor (CCII), DC biased by two currents Io.

By using the equivalent schematic presented in Fig. 2, which gives the input and output impedance, the behavior of the CC can be described by a voltage transfer and a current transfer as follows:

$$V_X = \alpha . V_Y \tag{1}$$

$$1_Z = \pm \beta . I_X \tag{2}$$

$$(\alpha, \beta \approx 1)$$

For this paper, only the second generation positive current conveyor (CCII +) is considered ($\beta = 1$). It could be demonstrated that α and β depend on the equivalent load at respectively Y, X and Z terminal. (R1 and C1, R2 and C2, R3 and C3 in the Fig. 2).

$$\alpha = \frac{R2}{R2 + R_X} \tag{3}$$

Table 1. Current conveyor applications.

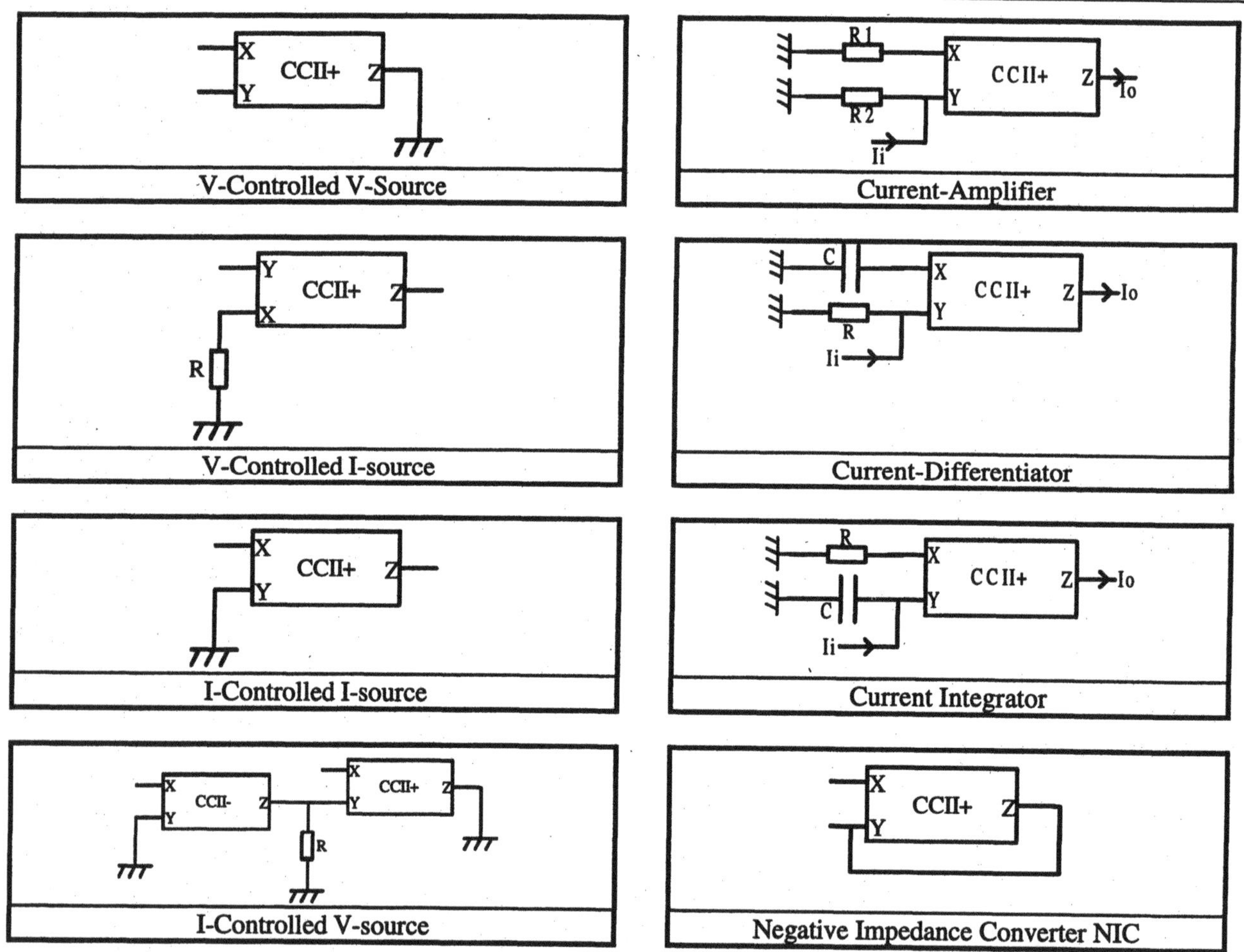

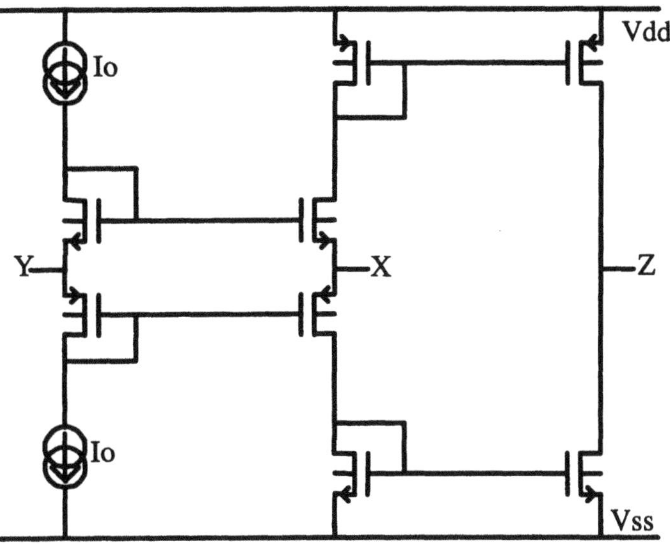

Fig. 1. Schematic of the current conveyor.

$$V_X = \frac{\alpha}{1 + s.\alpha.R_X.(C2 + C_X)}.V_Y \tag{4}$$

$$\beta = \frac{R_Z}{R_Z + R3} \tag{5}$$

$$I_Z = \frac{\beta}{1 + s.\beta.R_Z.(C3 + C_Z)}.I_X \tag{6}$$

An additional parameter is introduced when the considered input signal is a current Iin.

$$\gamma = \frac{R_Y}{R_Y + R1} \tag{7}$$

$$V_Y = \frac{\gamma.R1}{1 + s.\gamma.R1.(C1 + C_Y).Iin} \tag{8}$$

The current conveyor has been implemented using the AMS 1,2 μm CMOS technology previously used in [4] and chosen for its stability as far as process parameters are concerned. The voltage transfer and the current transfer are presented in Figs. 3 and 4. Table 2 summarizes CCs main performance characteristics. Frequency response for the current and voltage transfer are presented in Figs. 5 and 6.

Current conveyors have two main interesting characteristics for analog design. Firstly, basic analog functions are simple as they require only one

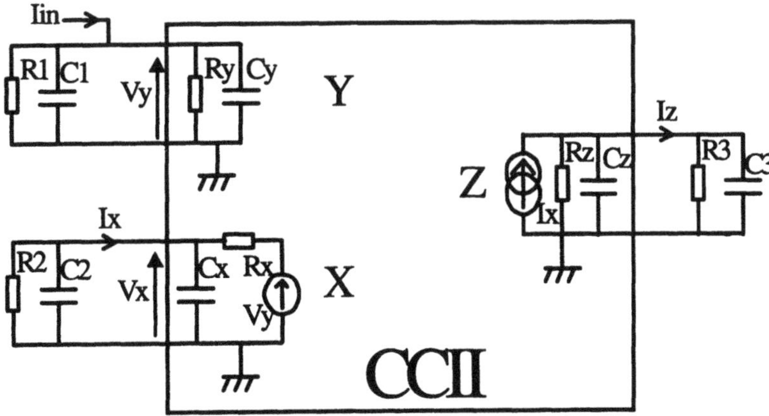

Fig. 2. Model schematic of the current conveyor.

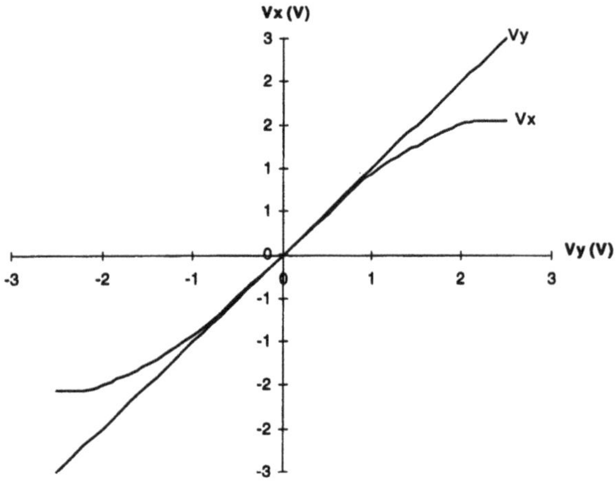

Fig. 3. Voltage transfer V_X/V_Y.

Table 2. CCs main performance characteristics.

Y Port Input Resistance	10 MΩ
X Port Input Resistance	977 Ω
Z Port Output Resistance	185 kΩ
Voltage Offset	100 μV
Current Offset	0,1 μA
– 3 db Frequency (RL = 1 mΩ, C = 1 pF)	3 MHz
Voltage Range	– 1,5 V to + 1,5 V
Current Range	– 200 μA to + 200 μA
Supply Voltage	– 2,5 V to + 2,5 V

or two discrete components (see Table 1). The proposed CC active area is 200 μm × 100 μm. Secondly, a large bandwidth (about 40 MHz for the considered process) can be achieved for high-frequency designs and complex applications [4].

3. Analog Elementary Cell

The proposed design for the Analog Elementary Cell (AEC) is based on current conveyors (CCs), programmable resistors and capacitors [5]. In this section, AEC design issues are discussed to show that CCs are well adapted for the implementation of an AEC. A key feature of the analog array concerns the area of each AEC. This area has to be as small as possible to allow the integration of a large number of cells and the implementation of complex analog functions. A CC is a suitable approach as it performs a wide range of analog functions with a low number of resistors and capacitors. Another key feature of the analog array is the interconnection architecture because analog circuits are very sensitive to the switching of signals. The major interest in the use of CCs is that this element may be used to perform an efficient interconnection between cells. Two generations of the cell are addressed in this section. The major difference is that the second generation elementary cell is a full-differential analog block.

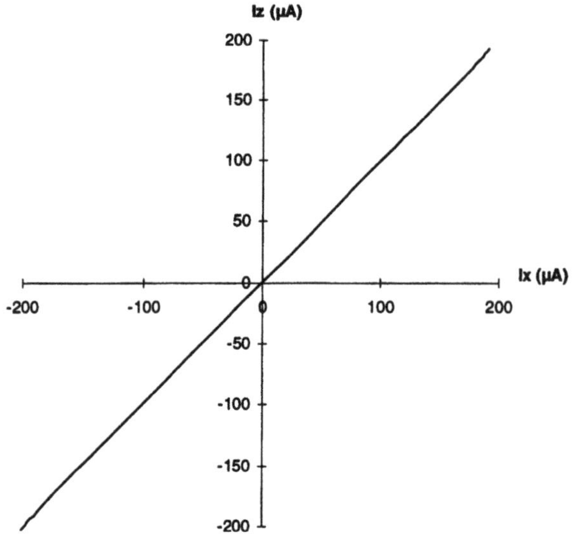

Fig. 4. Current transfer I_Z/I_X.

3.1. First Generation Elementary Cell

The design of an analog processing application is performed by cascading two CCs as shown in Fig. 7. The output port Z of the first CC is connected to the input port Y of the second CC. If several CCs port Z are connected to the same node, then the output currents are added and converted in voltage with a resistor before going to the next stage. This feature has influenced the design of the analog elementary cell presented in Fig. 8. This block uses only positive CCs (see previous section).

By changing the values of resistors and capacitors, this implementation can provide designers with programmable signal processing functions (see Table 1). This two-current-conveyors cell implements a non-ideal gyrator which exhibits simultaneously the

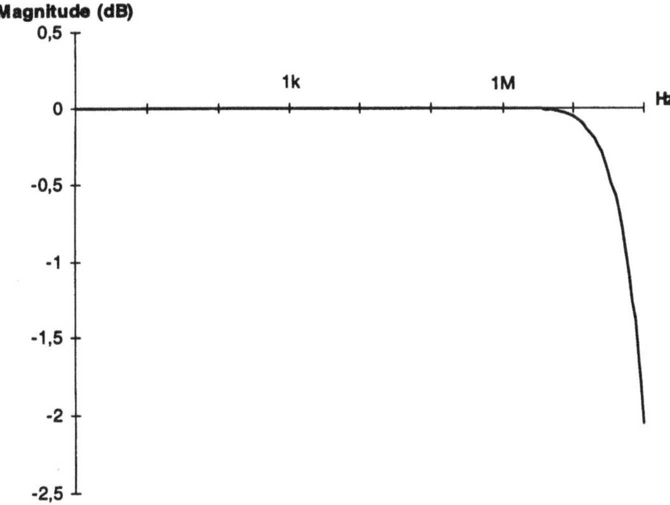

Fig. 5. Voltage transfer V_X/V_Y frequency response.

lowpass, bandpass and high-pass transfers. This block has been thoroughly studied in [6]. The interconnection aspect has to be considered at the beginning of the AEC design in order to be efficient because it influences the performance and the routability of the analog array. A CC can perform interconnection between cells. It simulates a pass switch or a non-pass switch by turning respectively on or off its two bias current sources [7]. The CC's equivalent input impedance is modified with the polarisation. The implementation of a switching element in an analog array requires a two way signal path device. The proposed cell based on a gyrator structure can be used as a two way analog signal switch (see Fig. 9). The performance of this switch are very similar to the ones of traditional integrated switches.

Many methods have been used to implement programmable resistors [8]. A CMOS resistor has

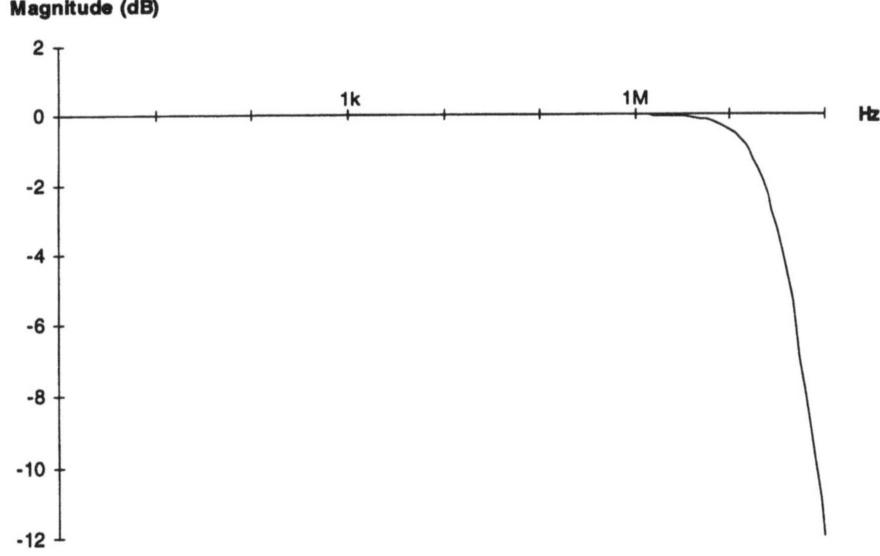

Fig. 6. Current transfer I_Z/I_X frequency response.

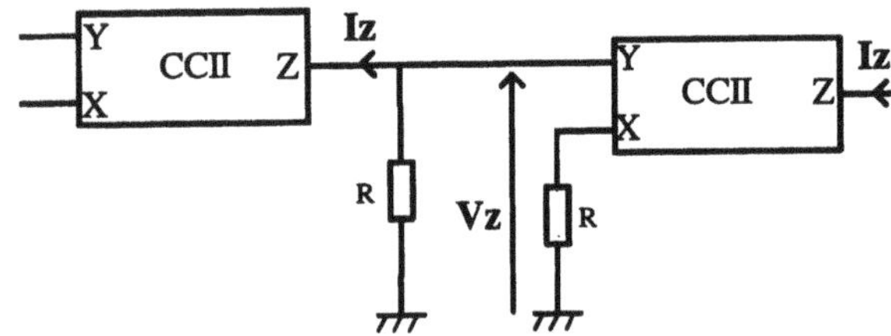

Fig. 7. CCs interconnections.

been used because the main idea was to have both good linearity and tuneability. The linearity of the resistor has to be as good as the linearity of the CC, otherwise the harmonic distortion component of the AEC will be dominated by the resistor nonlinearity. Besides, the tuning of the resistor requires a control circuit. Details on the setting of the programmable resistor and capacitor values can be found in Section 5. For the implementation of full CMOS resistors, OTA (Operational transconductance amplifier) in a unity feedback configuration has been often used [9,10]. Other approaches use transistors in linear region [11] or saturated region [12]. Because of low supply-voltages, e.g. $+/-$ 2,5 Volts, a resistor (see Fig. 10) based on CMOS transistors biased in linear region has been implemented whose equivalent value is:

$$R = \frac{1}{2.\mu.C_{OX}.\frac{W}{L}.(V_C - V_T)} \tag{9}$$

Parameters μ and C_{OX} are respectively the average carrier mobility in the channel and the gate oxide capacitance per unit area. L and W are respectively the length and the width of the transistor. V_T is the threshold voltage. V_C is the control voltage. The two transistors should be matched. Relation (9) is true for $V < (V_C - V_T)$. However V_T depends on $V_{BS} = V_B - V_S$. The potential of the bulk V_B is tied

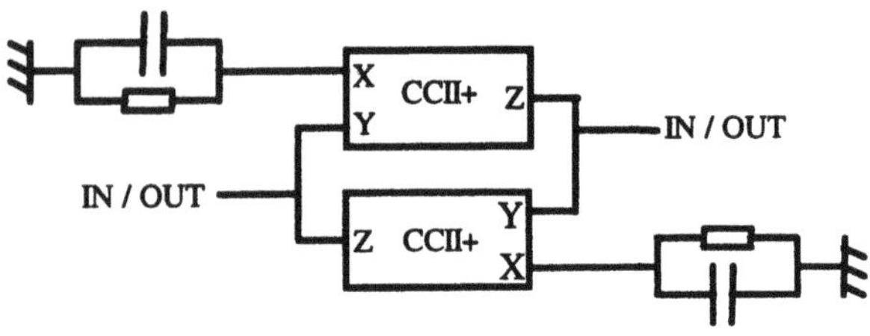

Fig. 8. Analog elementary cell.

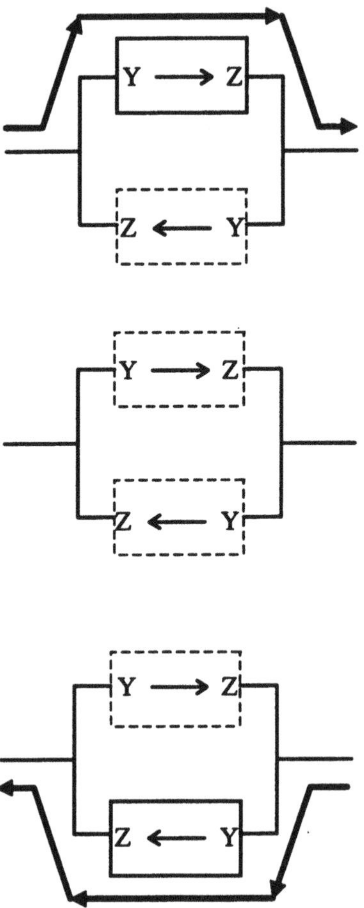

Fig. 9. Behavior of the analog switch.

to V_{SS}, but the potential V_S, for non-grounded application, is floating and then will affect the linearity of the resistor. Programming of the resistor value is achieved by changing the control voltage V_C. Fig. 11 shows the transfer characteristic current versus voltage and the Fig. 12 presents the value of the resistance versus frequency.

Two tuneable resistors are presented with two different scales W/L to obtain two overlapping ranges of values (65 kΩ to 120 kΩ and 100 kΩ to 350 kΩ). By connecting in a parallel association six tuneable resistors, with overlaping ranging values, a wide range of resistance can be achieved (from 500 Ω to 350 kΩ). The proposed resistor has been chosen because of its low active area (60 μm × 100 μm for a typical CMOS 1,2 μm process) but has no particular improvements compared to the available implementations. The programmable resistor can be set to an infinite

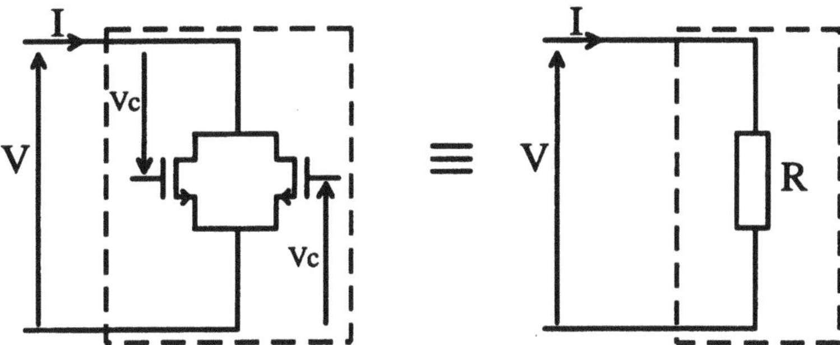

Fig. 10. Programmable resistor.

value. If the control voltage applied to the transistors gate of the resistor is set to V_{SS}, (the lower power-supply voltage), then the two transistors are in the cut-off region. The programmable resistor is withdrawn from the circuit. A Programmable capacitor is also required for the implementation of the AEC. Capacitor arrays have been previously designed [13]. Switches are used to set the capacitor value or to emulate resistors for switched-capacitor circuits. Because of the resistance of the non-ideal switches, the programmable capacitor appears to be non-linear. Besides, an important number of capacitors and switches is required to achieve a good tuneability which increases the area of the circuit. A CC based design has been developed and is presented in Fig. 13. The proposed programmable capacitor is based on

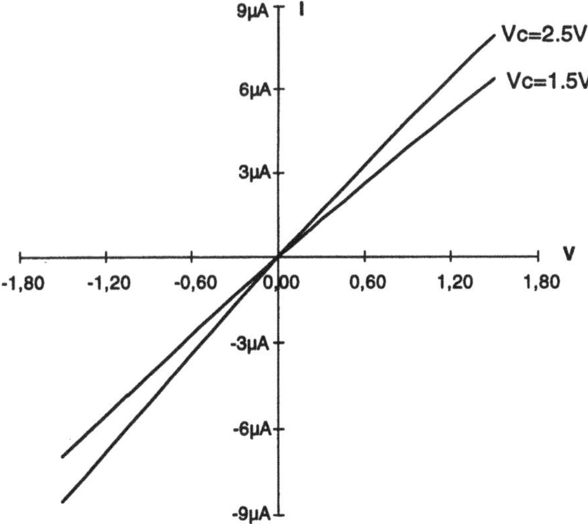

Fig. 11. Programmable resistor: transfer characteristic.

only one real capacitor. This technique seems to be more area-efficient than the capacitor array approach. One other key feature is the tuneability. The programmable capacitor is changed by a gain which depends on the value of a resistor. This resistor is controlled by a voltage and allows a good tuneability of the programmable capacitor (see the following results).

The programmable capacitor is based on the capacitive multiplier principle. The equivalent input impedance is:

$$Z_i = \frac{V_i}{I_i} = \frac{1}{s.\alpha.C} \tag{10}$$

The gain α is R_1/R_2. By changing the resistor R_1, the programmable capacitor is tuned from 10 pF to 4 nF (see Fig. 14), with a 180 μm × 200 μm active area for a typical CMOS 1,2 μm process. The frequency response of a low-pass first order filter using the programmable capacitor based on a real 10 pF capacitor and a 10 KΩ resistor is presented in Fig. 15.

The programmable capacitor can also be set to zero. If the current-conveyor polarisation is turned off then the equivalent capacitor will be set to zero and it is electrically withdrawn from the circuit.

3.2. *Second Generation Elementary Cell*

Several approaches [14,15] in the area of Field Programmable Analog Arrays have proposed full-differential circuits. In many industrial applications, widely used in sensors for example, or for signal processing functions, full-differential structure are required. A second generation elementary cell based

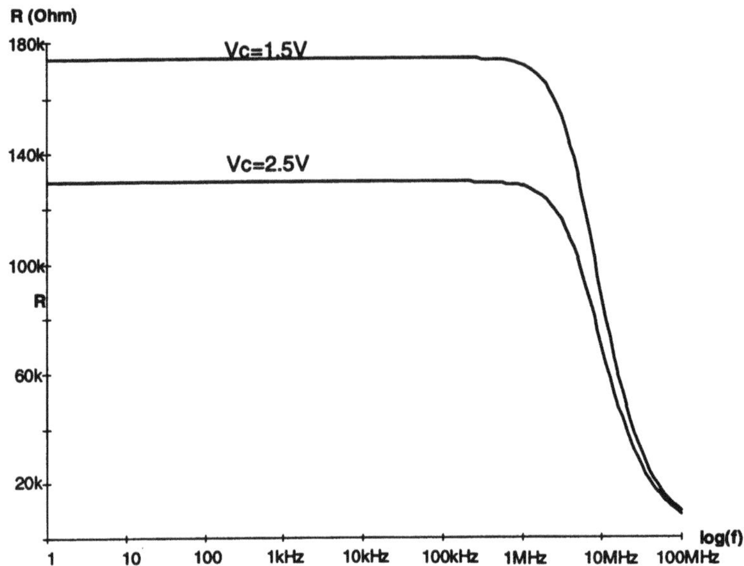

Fig. 12. Programmable resistor: R versus frequency.

on CCs, has been designed to propose a full-differential block. The major innovation is the use of a four MOSFET transconductor, introduced in [13,14] (see Fig. 16).

This realization is a true floating resistor. Assuming that all the devices should be matched and in ohmic region, an expression for the ac resistor can be obtained and the equivalent resistor can be found as:

$$R = \frac{V_1 - V_2}{I_1 - I_2} = \frac{1}{\mu . C_{OX} . \frac{W}{L} . (V_{C1} - V_{C2})} \tag{11}$$

Relation (11) is true if the conditions $V_1, V_2 < \min[V_{C1} - V_T$ - $V_{C2} - V_T]$ are verified. V_{C1} and V_{C2} are the control-voltages. The novel block, presented in Fig. 17, performs a larger set of functions than the first generation cell. The new circuit can behave as presented in Table 3. The next section presents some application examples like analog multiplication and AM modulation, which demonstrate the potentiality of this cell for the implementation of an analog array.

The full-differential block transfer function is:

$$\frac{V_{I+} - V_{I-}}{V_{O+} - V_{O-}} = -R_2 . \beta . \frac{1}{1 + sR_2C_2} . (V_{C1} - V_{C2}) \tag{12}$$

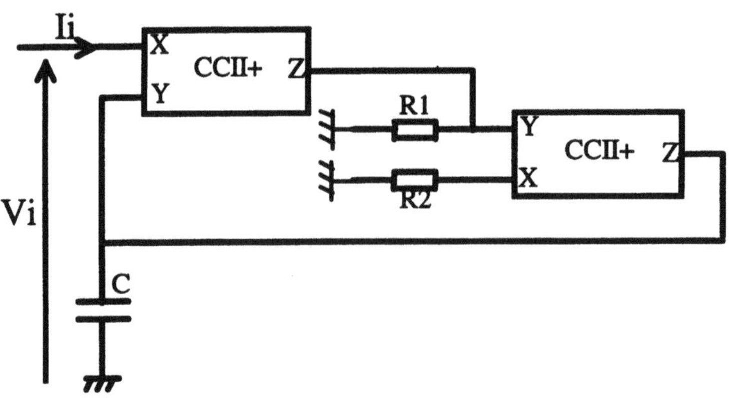

Fig. 13. Programmable capacitor.

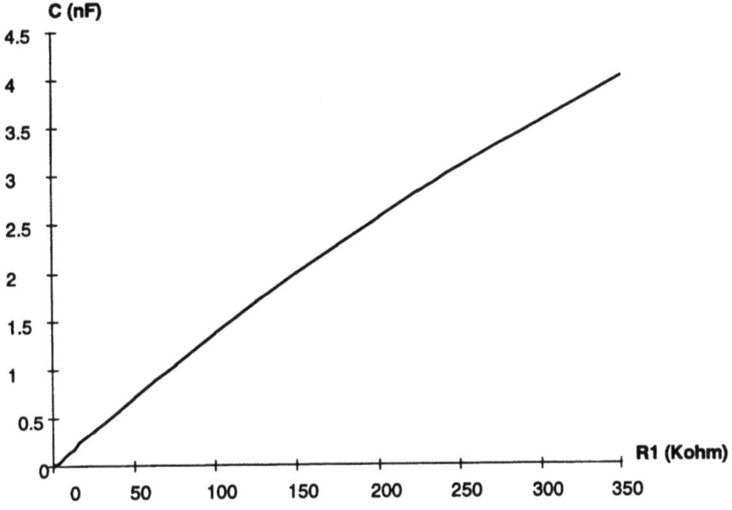

Fig. 14. Programmable capacitor: C versus R_1.

The parameter β corresponds to $\mu.C_{OX}.W/L$. (W and L are respectively the width and length of the transistor of the four MOSFET transconductor.) The resistor R_2 and capacitor C_2 are the programmable elements introduced in the previous subsection. The four MOSFET transconductor introduces the voltage difference $(V_{C1} - V_{C2})$ in the transfer function. This technique allows the implementation of an analog multiplier. In order to implement a two way signal path for the analog array, the gyrator structure is also used as presented in Fig. 18.

4. Elementary Cell Based Applications

Four application examples are described to show the real potentialities of the proposed AEC for the implementation of an analog array. Several basic analog applications have been studied in order to find what number of AECs were required and what kind of architecture was the best adapted for the array. The result of these studies has permitted the design of the proposed AEC based FPAA, presented in the next section.

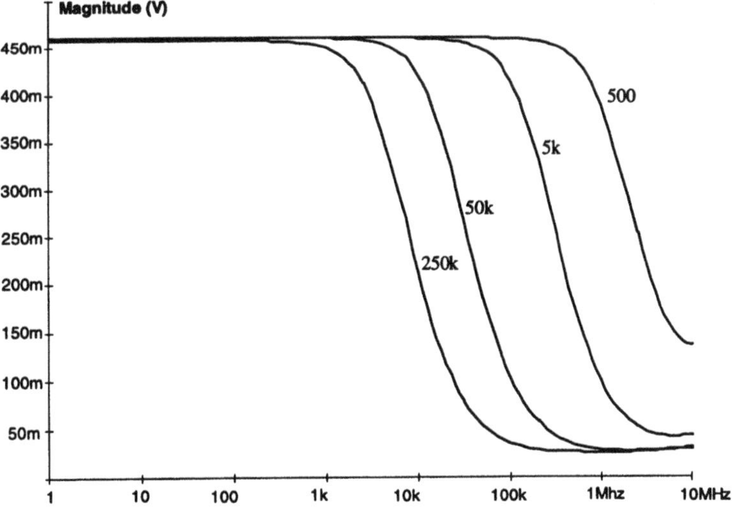

Fig. 15. First order filter (R_1 varying from 500 Ω to 250 kΩ).

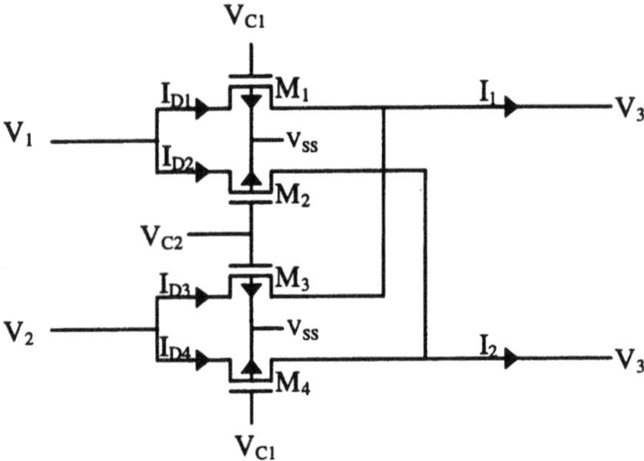

Fig. 16. Four MOSFET transconductor.

Table 3. Functionnalities of the circuit.

Non-inverter Amplifier
Inverter Amplifier
1st Order Low-pass Filter
Analog Multiplier
Comparator

4.1. Active Filter

A large number of active-filter realization have been proposed in litterature. Many of these can be implemented using current conveyors. A method, named ''adjoint networks'' [1], can be used to transform a voltage-mode circuit in a current-mode implementation.

Current conveyor based applications in the synthesis of various active filter transfer functions have received considerable attention [16,17]. Many of these designs suffer from low input impedance and voltage gain which can not be controlled independently of the other filter parameters (quality factor, cut-off frequency, etc). A universal active-filter network using current conveyors from which any type of filter characteristic can be realised, has been introduced in [18] by Toumazou and Lidgey. The proposed implementation uses each current conveyor as a voltage-to current converter, with a transconductance defined by a grounded resistor (see Fig. 19). Grounded capacitors, which are attractive for integrated circuit implementation (see precious section), are also used.

Circuit analysis yields to the following voltage transfer function:

$$\frac{V_o}{V_i} = \frac{s^2 C_1 C_2 R_4 \frac{R_5}{R_3} + s C_1 \frac{R_4}{R_2} + \frac{1}{R_1}}{s^2 C_1 C_2 R_4 \frac{R_5}{R_3'} + s C_1 \frac{R_4}{R_2'} + \frac{1}{R_1'}} \tag{13}$$

Highpass, lowpass bandpass and bandnotch transfer functions can be performed by the network by changing the resistor values of the previous relation. Although the filter complexity is important, the circuit offers several advantages such as high input impedance, control of characteristic frequency, quality factor and gain with grounded resistors and capacitors.

Another approach for the implementation of active filter concerns the use of non-ideal gyrator. This technique, introduced in [6], can be used to design active filter using two current conveyors, grounded resistors and capacitors. A non-ideal gyrator (see Fig. 20),

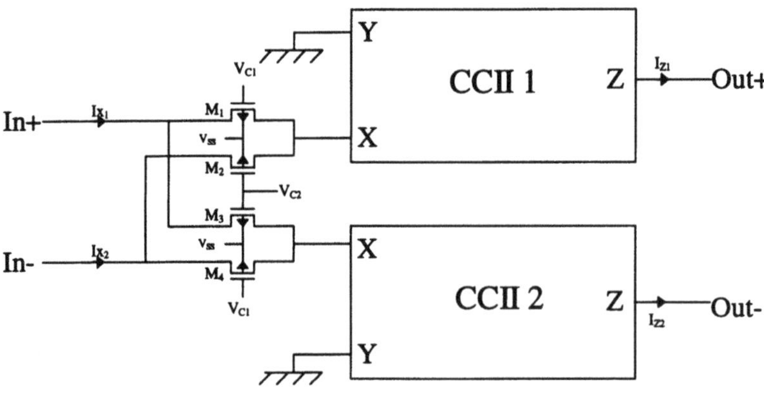

Fig. 17. Full-differential block.

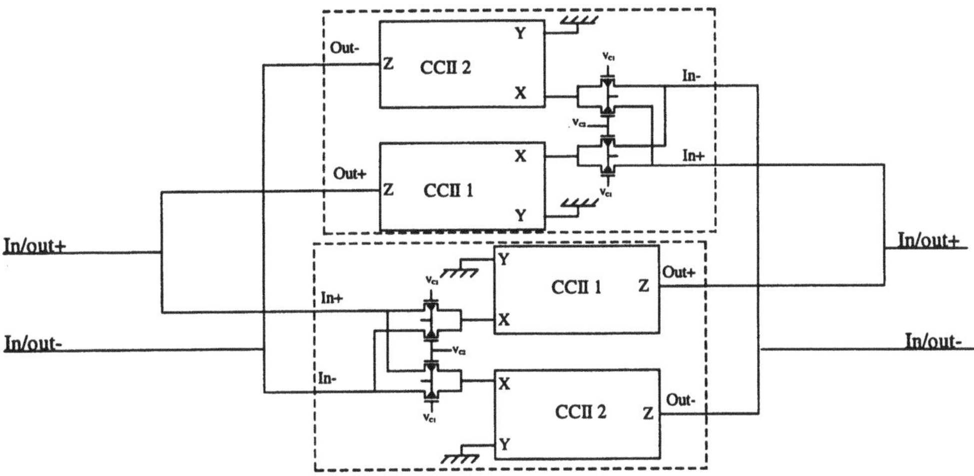

Fig. 18. Second generation analog elementary cell.

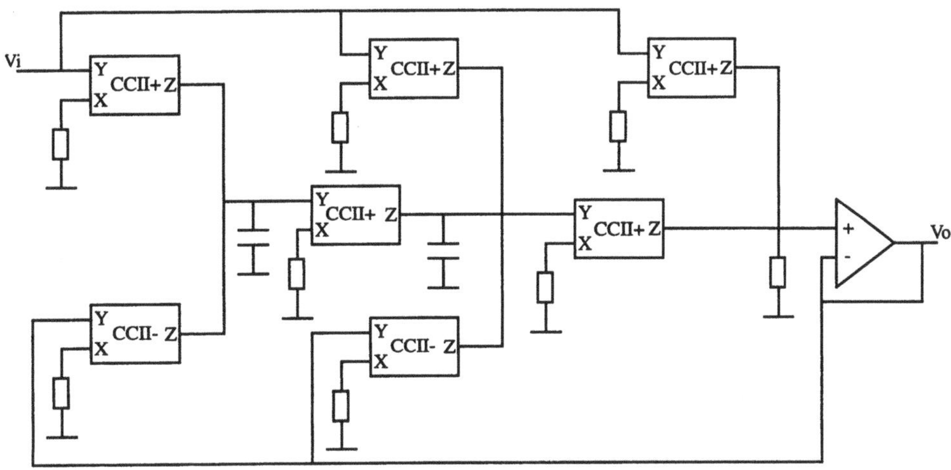

Fig. 19. Universal active-filter network.

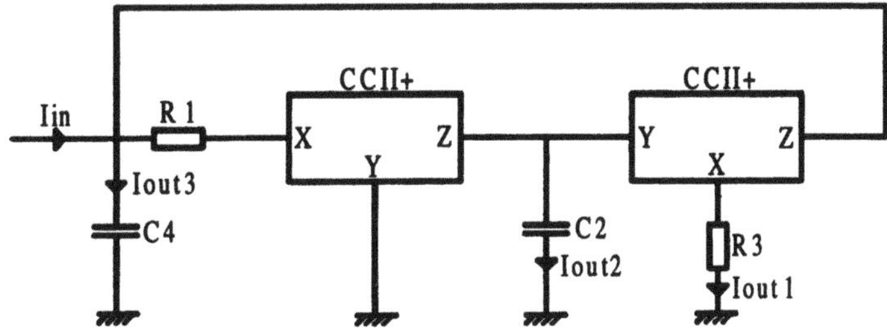

Fig. 20. Non-ideal gyrator.

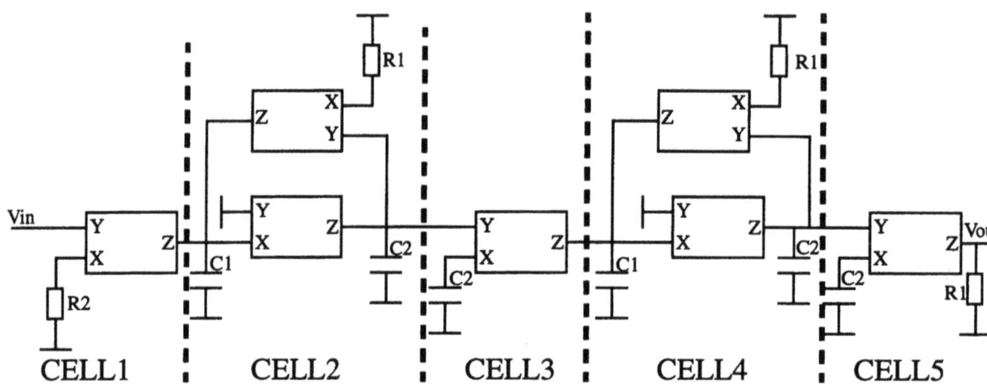

Fig. 21. 4th order filter.

exhibits simultaneously the lowpass, bandpass and highpass transfer with unity gain (see relations (14), (15) and (16)).

$$\frac{I_{out1}}{I_{in}} = \frac{1}{1 + sR_3C_2 + s^2R_1R_3C_2C_4} \tag{14}$$

$$\frac{I_{out2}}{I_{in}} = \frac{sR_3C_2}{1 + sR_3C_2 + s^2R_1R_3C_2C_4} \tag{15}$$

$$\frac{I_{out3}}{I_{in}} = \frac{s^2R_1R_3C_2C_4}{1 + sR_3C_2 + s^2R_1R_3C_2C_4} \tag{16}$$

The cut-off frequency is:

$$f_O = \frac{1}{2\pi\sqrt{R_1R_3C_2C_4}} \tag{17}$$

A circuit based on five first generation AECs is described for the implementation of a 4th order filter. The proposed circuit of Fig. 21 consists of two AECs (cell 2 and cell 4), used as non-ideal gyrators, two AECs (cell 1 and 3) as voltage-to-current converters and one AEC (cell 5) as a current-to-voltage converter. Fig. 22 shows the frequency response.

The resistor R_1 which appears in the non-ideal gyrator structure, corresponds to the input resistor R_X of the current conveyor (see Section 2). The following values were used (in correspondance with Fig. 20): $R_1 = 10\,\text{k}\Omega$, $R_2 = 500\,\Omega$, $C_1 = 100\,\text{pF}$ and $C_2 = 600\,\text{pF}$. The theoretical value for the cut-off frequency f_O is 207 kHz.

The proposed AEC appears to suit very-well for the implementation of such active filters. Grounded resistors and capacitors of the AEC are used to change transfer function parameters (gain, quality

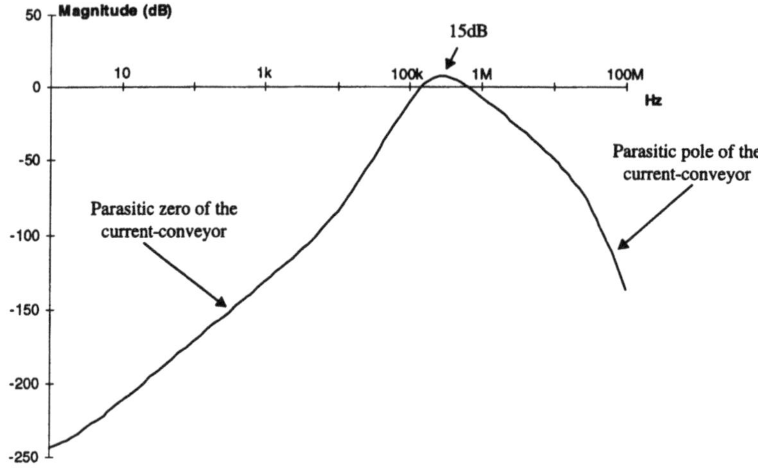

Fig. 22. 4th order filter frequency response.

factor, cut-off frequency, etc). The analog array, presented in the next section, allows the implementation of the universal active-filter network that has been described. Moreover, the implementation of local 2nd order filters can be performed by the non-ideal gyrator structure of the AEC.

4.2. *Analog Multiplier*

Analog multiplier are key building blocks in adaptive equalization, frequency translation, waveform generation and other signal processing circuits. The proposed second generation AEC performs analog multiplication using the four MOSFET transconductor (see Section 3). The differential input signal is applied to the input of the AEC. The second signal is used to set the control voltages of the four MOSFET transconductor. Fig. 23 presents the multiplication of two sinusoidal signal which have the same frequency. The multiplier behaves as a frequency doubler.

Fig. 24 shows the AEC being used for AM modulation with a 10 kHz sinusoidal input signal and a 1 kHz sinusoidal signal to control the four MOSFET transconductor. The configuration of the AEC is the same as the multiplier one. A DC voltage is introduced in the control voltages of the four MOSFET transconductor in order to prevent a zero value of the voltage difference $(V_{C1} - V_{C2})$.

4.3. *Rectifier*

A full-wave rectifier has been designed using two AECs. The first AEC is used as a comparator to change the control voltages of the four MOSFET transconductor of the second AEC. This second AEC is used to change the polarity of the input signal in order to invert this signal for each half period. Results for the rectifier are shown in Fig. 25.

5. Analog Array

In this section, analog array design issues are discussed to exhibit the real potentialities of the proposed current conveyor based AEC for the implementation of a FPAA. Special attention for the design of analog applications has to be paid to several parameters like noise level, distortion, dynamic range,

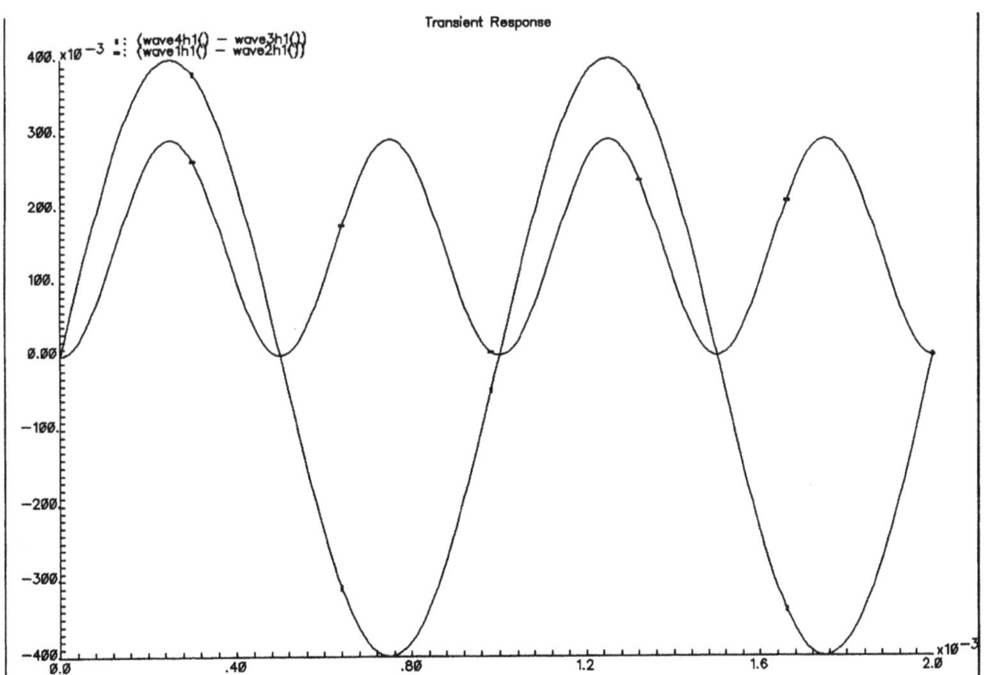

Fig. 23. Frequency doubler.

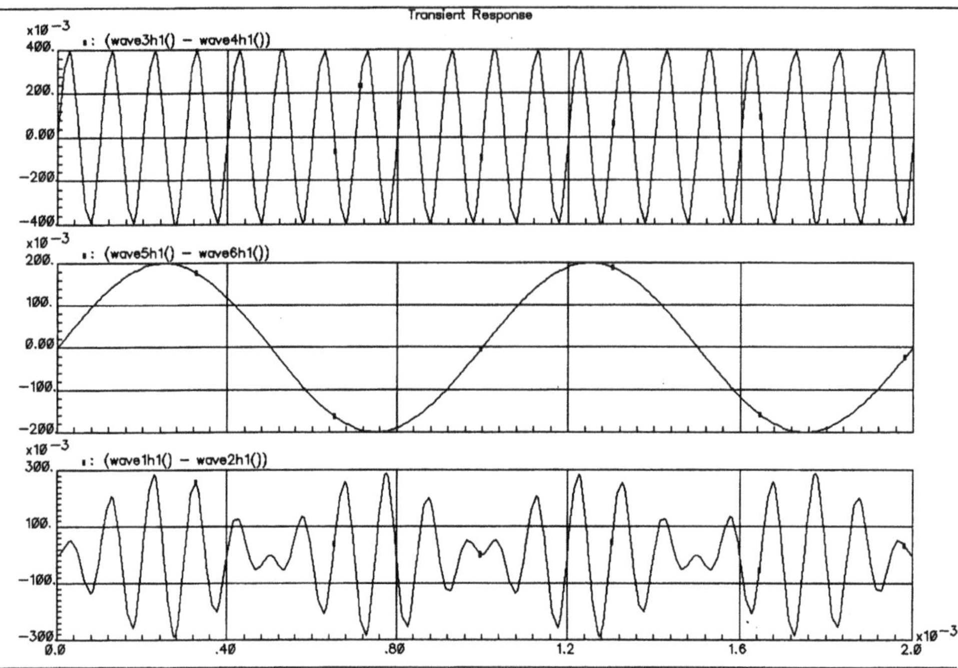

Fig. 24. AM modulation.

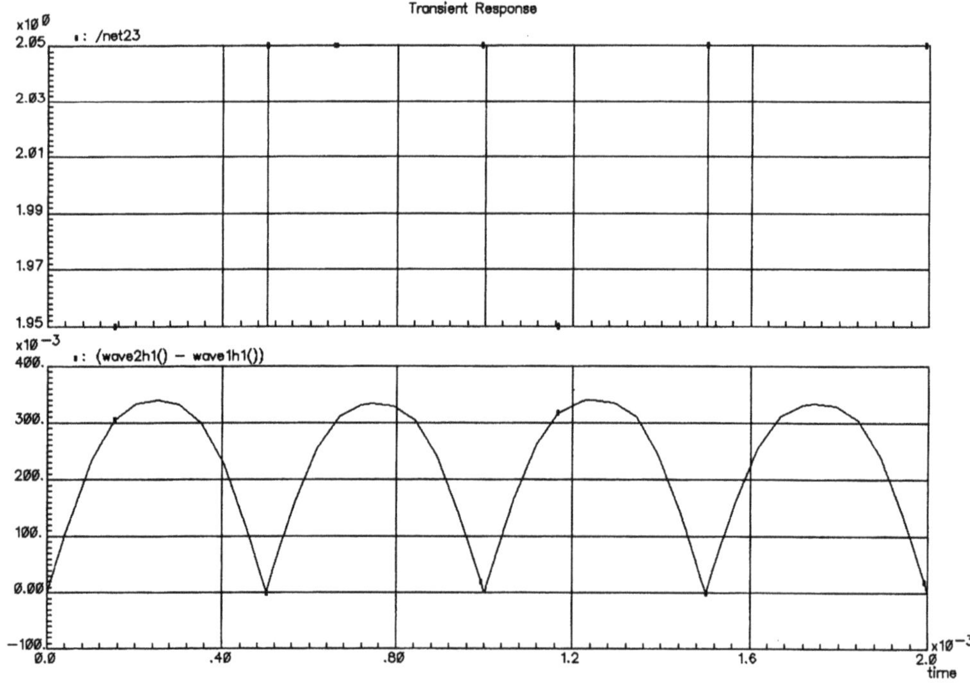

Fig. 25. Full-wave rectifier.

etc, but a key feature is that analog applications can be switched-capacitor or continuous-time based designs. These two techniques do not offer the same trades-off between performance and tuneability. The switched-capacitor approach is digitaly controlled and provide a good tuneability for the circuit characteristics. One of the main drawback of this technique is that the input signal frequency band is limited because of the Shannon theorem. Another problem concerns the important area dedicated to the control. Continuous-time operation does not allow to achieve a high range of tuneability, but is not band limited. Besides, design techniques and specific technologies, like BiCMOS or bipolar technologies, can be used to implement high-performance and complex analog applications. The continuous-time technique for the implementation of FPAAs, has been introduced by two research approaches.

Gulak and Lee [14–19] have proposed an approach based on a modified MOSFET transconductor (see Fig. 26) used both as switch to interconnect configurable analog blocks (CABs) and as programmable resistor. The function of each CAB and the connections among CABs are programmed by the content of shift registers which activate the switches in the interconnection network. The CABs containing the OP-AMPs also contain switched feedback capacitor programmed as integrator, OP-AMP or comparator. The use of four-transistor switches for connection between CABs achieves high-linearity, low-power dissipation and high-flexibility for a wide range of applications. But the performance of the circuit is degraded due to finite resistance of switches and the limitation of the frequency by the Op-Amps, thus high-frequency applications are not possible.

Another approach, developed by Pierzchala and Perkowski [15–21], proposes a device based on a regular, square array of processing cells, interconnected with a local and a global interconnection pattern. The cell core comprises the Gilbert multiplier, a wide-band current-mode amplifier and some additional circuitry is used for programming. Each cell is connected to its four nearest neighbors by a two-way current-mode signal interconnection which allows the local interconnection without the use of switch. The cell processes current-mode, differential signals to implement mathematical functions (see Fig. 27). X_i are inputs of the cell and Y the result. W_i and W_j are programmable weights. The result Y can be integrated with a programmable loss-factor, and

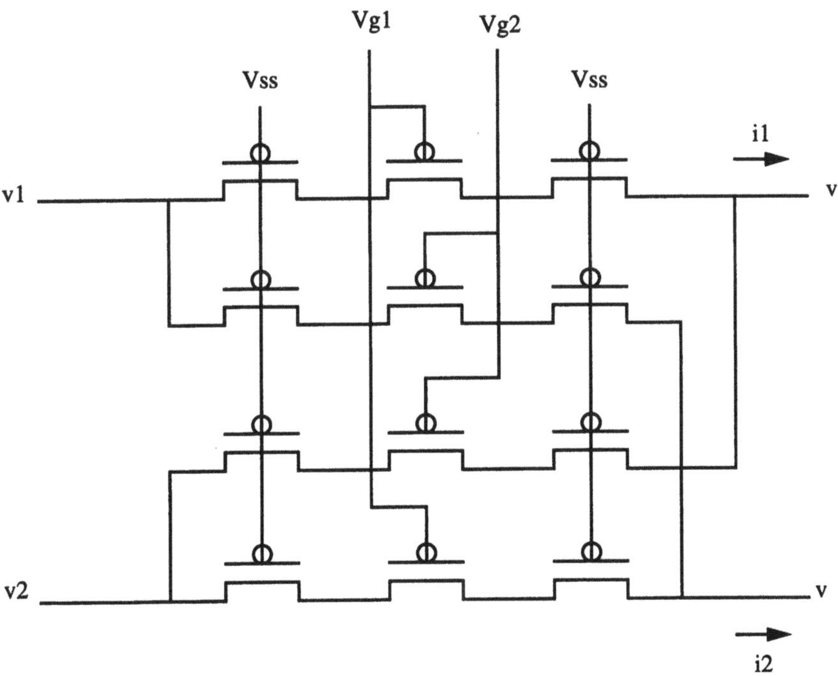

Fig. 26. Modified MOSFET transconductor used by Gulak and Lee.

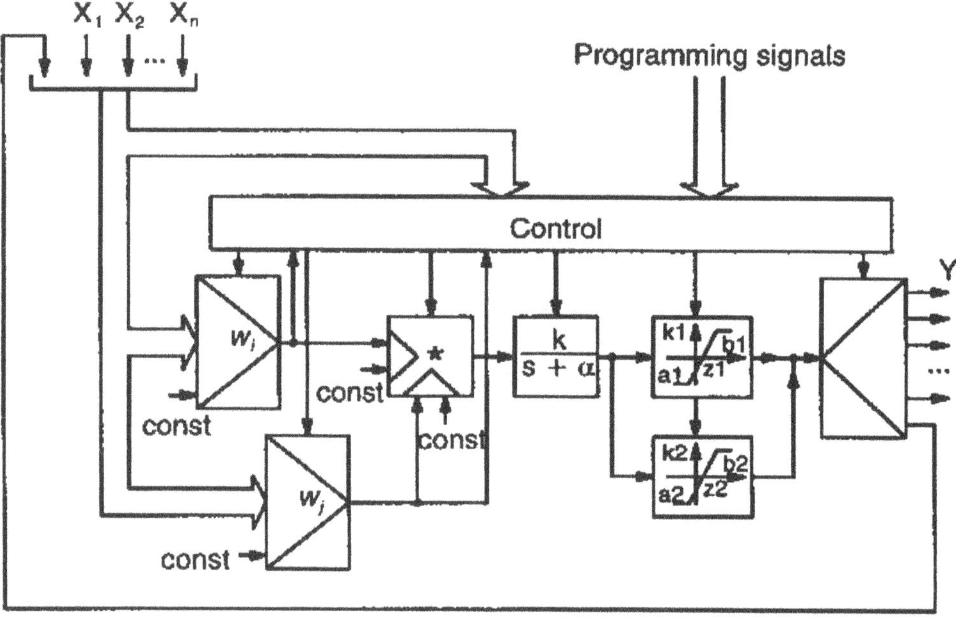

Fig. 27. Block diagram of the cell developed by Pierzchala and Perkowski.

passed through a programmable non-linear thresholding and clipping circuit.

The proposed analog elementary cell presented in Section 3, achieves feedback loops and offers a possible connection between nearest neighbors by a two-way signal interconnection. The FPAA is based on a regular pattern, rows and columns being implemented by cascading AECs. Each node, which is shared by four AECs, consists in a parallel association of a programmable resistor and a programmable capacitor. If port Z of many AECs are connected to the same node, then the output currents are added and converted in voltage by the resistor of the node before going to the next stage (see Fig. 28). This technique can be used for the two generations of AEC.

The topology presented in this section is based on local interconnections. If a signal must be transfered from a node to a remote node, it will use a path which consists of many AECs. A second interconnection scheme could be superimposed to the first one. Some remote nodes could be directly connected by one AEC. This technique would increase the flexibility of the analog array but also its complexity. Using the proposed topology, an analog application is implemented programming the AECs of the circuit. The transfer function is divided into functional blocks which correspond to well-programmed AECs. The function of an AEC is set by two different control systems. The first one, which is digital, is used to set CCs, programmable resistors and capacitors active or not. The CCs are biased by two current sources (see Section 2) which are MOS transistors controlled by reference voltages applied on their gate. If the gate voltage is set to V_{SS} or V_{DD}, respectively the negative and positive power supply voltage, then the two transistors are in cut-off region and the polarisation of the CC is turned off (see Fig. 29), which behaves as a non-pass switch or as an active element.

Besides, the programmable resistor and capacitor values are set by control voltages. The same technique used for the control of the CCs polarization, is implemented to set resistors and capacitors active or not. Then, a serial bit register implements the control systems allowing an element, a CC, a programmable resistor or capacitor, to be active or not. A second control system, which is analog, is used to set the programmable resistor and capacitor values. Two types of configuration can be used. Digital registers which are converted with a digital to analog converter to provide control voltages, can be used to store the resistor and capacitors values. The control voltage is locally stored, near the programmable element, on a capacitor (see Fig. 30), and needs to be periodically

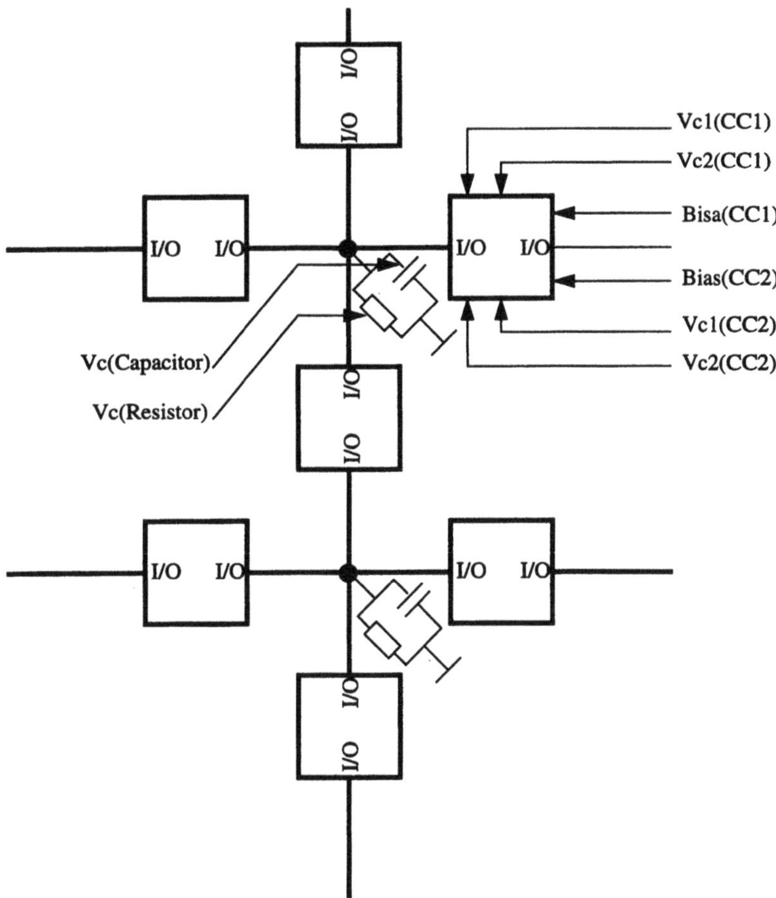

Fig. 28. Schematic of the AEC-based analog array.

refreshed. The second type of configuration consists in using a non-volatile analog memory. This technique has been thoroughly studied in [22,23].

6. Conclusion

A new approach for designing FPAAs has been described. Its major feature is the use of current conveyors to achieve a wide range of analog functions such as amplification, comparison, filtering, multiplication, rectification, etc. A current conveyor based analog block has been introduced for the implementation of FPAAs. This particular circuit performs analog signal processing functions and is also used as a switching element. The proposed approach permits the design of an interconnection architecture based on a two-way signal path elementary cell, without using MOSFET based switches. Programming features are addressed through the description of programmable resistor and capacitor.

Future work will consist in validating the proposed approach with a physical implementation of a Field Programmable Analog Array based on Current Conveyors.

Acknowledgments

The authors wish to thank Edmund Pierzchala, Bruno Allard, Stéphane Cattet, Béatrice Françon and the reviewers for their helpful comments and suggestions during the preparation of this paper.

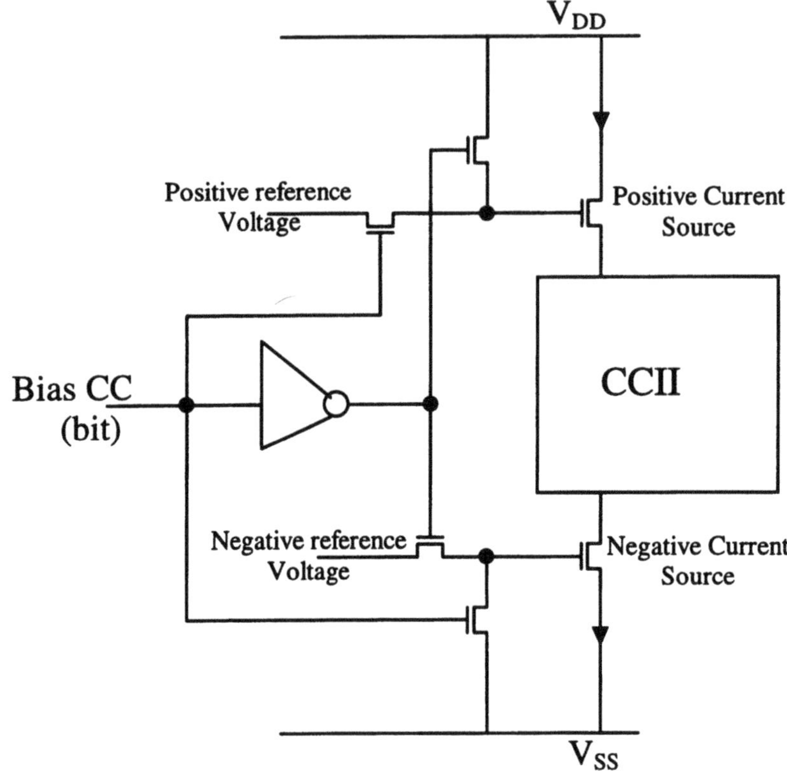

Fig. 29. Control of the CCs polarization.

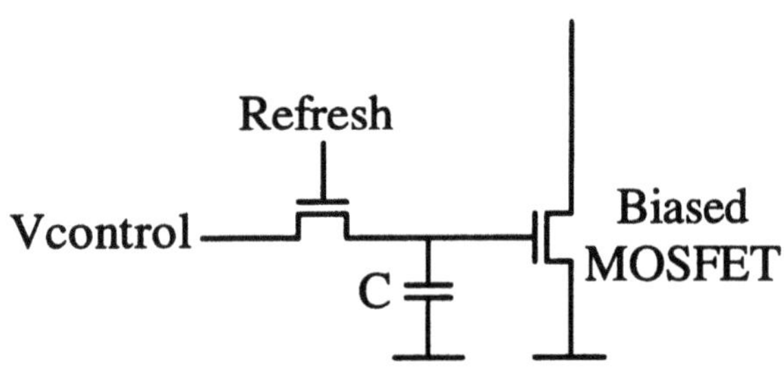

Fig. 30. Local storing of control voltage.

References

1. C. Toumazou, F. J. Lidgey, and D. G. Haigh, "Analog IC design: the current-mode approach." Chapter 3, Current Conveyor Theory and Practice, London Peregrinus, 1990.
2. A. S. Sedra, C. W. Roberts, and F. Gohh, "The Current-conveyors: history, progress and new results." *IEE Proceedings* 137G(2), April 1990.
3. B. Wilson, "Recent developments in current-conveyors and current-mode circuits." *IEE Proceedings* 137G(2), April 1990.
4. F. Bergouignan, N. Abouchi, R. Grisel, G. Caille, and J. Caravana, "Designs of a Logarithmic and Exponential Amplifier Using Current Conveyors." *Third IEEE International Conference on Electronics, Circuits and Systems (ICECS''96)*, Rodos, GREECE, pp. 61–62, Oct. 1996.
5. C. Premont, R. Grisel, N. Abouchi, and J. P. Chante, "A Current Conveyor based Field Programmable Analog Array." *MWSCAS'96 Proc.*, Ames, Iowa, USA, Aug. 1996, pp. 155–157.
6. A. Fabre and M. Alami, "Universal Current Mode Biquad Implemented From Second Generation Current Conveyors." *IEEE Trans. on Circuits and Systems I* 42(7), pp. 383–385, July 1995.
7. A. Fabre et al., "High-Frenquency Applications based on Current Conveyors." *IEEE Trans. on Circuits and Systems I* 43, pp. 82–91, Feb. 1996.
8. J. Silva-Martinez, M. Steyaert, and W. Sansen, "High-Performance CMOS Continuous-time filters." Kluwer Academic Publishers, 1993.
9. K. Nagaraj, "New CMOS Floating Voltage-Controlled resistor." *IEE Electronics Letters* 22, pp. 667–668, June 1986.
10. Y. Tsividis and P. Antognetti, "Design of MOS VLSI Circuits for telecommunications." Prentice-Hall, Inc., Englewood Cliffs, New Jersey, 1985.
11. M. Banu and Y. Tsividis, "Floating Voltage Controlled resistors in CMOS technology." *IEE Electronics Letters* 18, pp. 678–679, July 1982.
12. S. P. Singh, J. V. Hanson, and J. Vlash, "A new floating resistor for CMOS technology." *IEEE Trans. on Circuits and Systems* CAS-36, pp. 1217–1220, Sept. 1989.

13. P. E. Allen, R. L. Geiger, and N. R. Strader, "VLSI Design of Analog and Digital Circuits." McGraw-Hill Int. Ed., Electronics Engineering Serie, 1990.
14. E. Lee and G. Gulak, "Field Programmable Analog Array based on Mosfet transconductors." *IEE Electronic Letters* 28(1), pp. 28–29, Jan. 1992
15. E. Pierzchala, M. A. Perkowski, and S. Grygiel, "A Field Programmable Analog Array for Continuous, Fuzzy and Multi-valued Logic Applications." *IEEE ISMVL, Boston, Mass., Proceedings*, May 1994.
16. A. Fabre and M. Alami, "Insensitive current-mode biquad implementation based translinear current conveyors." *IEEE Proc. EUROASIC''92*, Paris, June 1992, pp. 126–130.
17. A. Fabre, F. Martin, and M. Hanafi, "Cuurent-mode allpass/notch and bandpass filters with reduced sensitivities." *IEE Electronics Letters* 26, pp. 1495–1496, Aug. 1990.
18. C. Toumazou and F. J. Lidgey, "Universal active filters using current conveyors." *IEE Electronics Letters* 22(12), pp. 662–664, June 1996.
19. E. Lee and G. Gulak, "A Transconductor based Field-Programmable Analog Array." *ISSCC Digest of Technical Papers*, pp. 198–199, Feb. 1995.
20. E. Pierzchala and M. A. Perkowski, "High-Speed Field-Programmable Analog Array Architecture design." *FPGA''94*, Berkeley, Calif., Feb. 1994.
21. E. Pierzchala, M. A. Perkowski, P. V. Hallen, and R. Schaumann, "Current-Mode Amplifier Integrator for a Field-Programmable Analog Array." *ISSCC Digest of Technical Papers*, pp. 196–197, Feb. 1995.
22. A. Thomsen and M. A. Brooke, "Low Control Voltage Programming of Floating Gate MOSFETs and Applications." *IEEE Trans. on Circuits and Systems* 41(6), pp. 443–451, June 1994.
23. A. Thomsen and M. A. Brooke, "A Floating gate MOSFET with tunneling injector fabricated using a standard double polysilicon CMOS process." *IEEE Electron. Device Lett.* 12(3), pp. 111–113, Mar. 1991.

Christophe Premont was born in Thionville, France, in 1971. He received his engineering degree in electronics in 1994 from the Institut de Chimie et de Physique Industrielle de Lyon (ICPI Lyon). His Ph.D. Research is concerned with the development of a Field Programmable Analog Array (FPAA). His research interests include VHDL modeling and analog circuit design.

Richard Grisel was born in Alizay, France, in 1959. He received his Ph.D. Degree in January 1987 from Rouen University and High Level Research Degree from St Etienne University. He is currently the head of the "Electronics and microelectronics system" research group at LISA EP CNRS 0092, at the Ecole Supérieure de Chimie et Physique of Lyon (CPE Lyon). His research field concerns methodology of specific analog circuits and study of architecture performances.

Nacer Abouchi was born in Setif, Algeria, in 1962. He received his Ph.D. Degree in July 1990 from the Institut National des Sciences Appliquees (INSA Lyon). His Ph.D. research was concerned with switching in networks (LAN and MAN). He is currently a professor in electronics and microelectronics. His research field concerns analog circuits and study of architecture performances.

Jean-Pierre Chante received the "Doctorat d''Etat" from the University of Lyon, France in

1981. From 1980 to 1986, he managed a research team in the field of power semiconductor devices at the Central School of Engineers of Lyon. Since 1986, he has been a Professor of Electronic Components and Applied Electronics at the National Institute of Applied Sciences (INSA) of Lyon, where he is the leader of the Power Devices and Applications team which is a part of the Electrical Engineering Center of Lyon (CEGELY). He is also in charge of the CIMIRLY which is a regional research center in the microelectronic field. His interests are in high-temperature electronics, SiC-based components, advanced power devices and CAD tools for power electronics.

Analog Integrated Circuits and Signal Processing, 17, 125–142 (1998)

A Current-Mode based Field-Programmable Analog Array for Signal Processing Applications

S. H. K. EMBABI, X. QUAN, N. OKI**, A. MANJREKAR AND E. SÁNCHEZ-SINENCIO

*Dept. of Electrical Engineering, Texas A&M University, College Station, TX 77843, USA, **Dept. of Electrical Engineering, Fac. de Eng. de Ilha Solteira - UNESP, Ilha Solteira, SP 15385-000, Brazil*

Received July 12, 1996; Accepted April 23, 1997

Abstract. This paper presents a new approach to develop Field Programmable Analog Arrays (FPAAs),[1] which avoids excessive number of programming elements in the signal path, thus enhancing the performance. The paper also introduces a novel FPAA architecture, devoid of the conventional switching and connection modules. The proposed FPAA is based on simple current mode sub-circuits. An uncompounded methodology has been employed for the programming of the Configurable Analog Cell (CAC). Current mode approach has enabled the operation of the FPAA presented here, over almost three decades of frequency range. We have demonstrated the feasibility of the FPAA by implementing some signal processing functions.

Key Words: Field-Programmable Analog Array, current-made circuits, continuous time filters

1. Introduction

Field Programmable Gate Arrays (FPGAs) are now widely accepted in the digital design community. They are attractive for prototyping and low volume products, and have emerged as the preeminent solution to the time-to-market problems. On the analog side, there have been very few attempts at developing analog arrays for semi-custom design [1–7]. Field Programmable Analog Arrays (FPAA) provide a very convenient medium in which analog circuits and systems can be designed and implemented in a very short time frame.

FPAA and FPGA are similar in that they both consist of a modular array of cells which can realize a number of functions, and that both of them would conventionally, need switching cells for configuration and routing. However, as will be explained in Section 2, the proposed architecture obviates the use of a switching cell. The similarities between the FPGA and FPAA offer the potential to integrate the two, to yield a Field Programmable Mixed Signal Array (FPMSA). A fully functional FPAA, with an associated CAD tool for its user friendly configuration, would make it possible, even for an individual with limited knowledge of analog circuits, to build semi-custom analog circuits with immense ease.

Corresponding Author: S. H. K. Embabi, Dept. of Electrical Engineering, Texas A&M University, College Station, TX 77843, USA, Telephone: (409) 845 7160, Email: embabi@eesun2.tamu.edu

Lack of definition for an analog primitive, that can be found in a large spectrum of analog circuits makes the task of developing a FPAA quite complicated. While the ultimate goal is to define a generic FPAA which would be capable of implementing almost any analog function, this may be rather difficult to realize. Instead, it is more pragmatic to categorize the analog circuits into different groups and define optimum circuit primitives for each, thus leading to a family of FPAAs. In that direction, switched current and switched capacitor based FPAAs may be utilized for the audio frequency range, whereas, OTA-C and current mode based FPAAs can be employed at higher frequencies. The FPAA presented in this paper is current mode based (see Section 3.1).

A striking difference between the FPGA and the FPAA is that, unlike the FPGA, reducing the parasitics introduced by the programming devices is more imperative in a FPAA. There have been some efforts made to circumvent the problems caused by parasitic resistance and capacitance of the switches. To do so, in the past, a four transistor transconductor has been used, where the parasitic capacitance in each transistor can be canceled by proper matching and differential signaling [1]. Yet another approach employs the existing switches of a switched capacitor topology [2]. Very few programming devices can be tolerated in the signal path. This has been a major concern in developing the FPAAs. Towards that, a unique architecture of the CAC has been put forth, which will be explained in details in Section 3.1. The proposed FPAA architecture and the CAC topology is a balanced compromise between flexibility and the number of programming switches in the signal path, which permits the hardwiring of most of the signal path. A number of issues which are fundamental to the FPAA such as the architectures, the structure and the primitives of the configurable analog arrays, the programming devices and their impact on the performance of the analog system and the interconnection networks, have been addressed. The FPAA boasts of such features as, high frequency operation, a simple programming methodology, operation over almost three decades of frequency, and use of standard digital CMOS fabrication process.

2. The Proposed FPAA Architecture

Granularity is an important aspect of any array based architecture. The correct trade-off between granularity and the performance needs to be struck. Lower granularity implies better flexibility and utilization of silicon area, but it also necessitates a more articulate connection and routing module. On the other hand, routing and the programming switches introduce parasitic resistances and capacitances, which degrade the performance of analog circuits. Therefore, the granularity of the FPAA cells must be large enough to avoid degradation due to parasitic resistances and capacitances [6,7]. This factor has strongly influenced the design of the CAC and the overall FPAA architecture. The result of above consideration is a FPAA comprising each CAC connected via switches, to only up to 8 neighboring CACs and to itself, without the need for a dedicated connection module. While there exists limited connectivity, this scheme offers the advantage of reduced number of switches in the signal path. This is the uniqueness when compared with a FPGA approach which has a switching and a connection module. Although the architecture allows lesser freedom of connectivity within the array, it should be noted that this is not a drawback, since most analog circuits are characterized by predominantly local interconnections [5].

The proposed FPAA architecture (see Fig. 1) consists of a modular array of cells. Each of the cells includes three sub-cells: a Configurable Analog Cell (CAC), a Programming Register (PR) and Programming Logic (PL).

A circuit topology for the CAC has been developed as will be discussed in Section 3.1. Its functionality can be programmed digitally using a function word (FW). The characteristics of the configured function can also be programmed through the digital Parameter

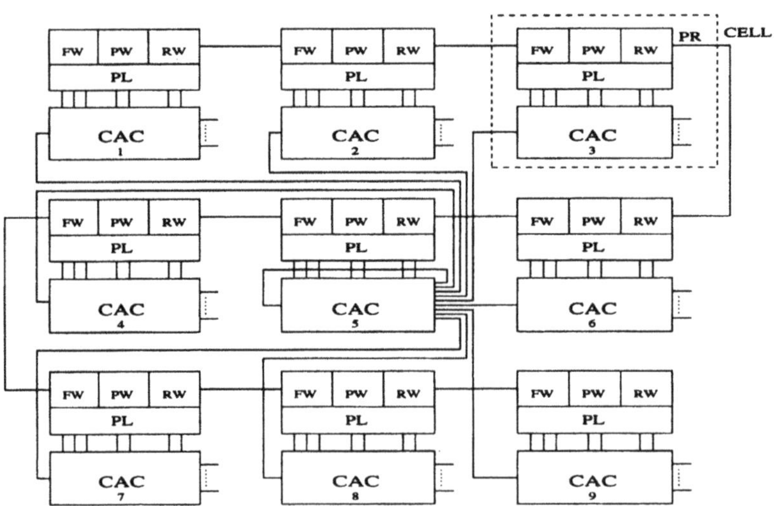

Fig. 1. Proposed architecture of the FPAA.

Word (PW). The desired connectivity between the CACs is configured through the Routing Word (RW). The output of each CAC is connected to its closest neighboring CACs (see the middle cell in Fig. 1). The digital control words for each CAC are held in the respective PR. All the PRs are connected in series and they together act as a single shift register which can be loaded with the desired programming bit stream through one external pin. This greatly, simplifies the programming process.

3. The Configurable Analog Cell

3.1. Architecture of the CAC

The commonly used functional blocks in signal processing systems (e.g. Filters and Data Converters) are integrator, amplifier, attenuator and comparator. Thus the goal in sight, while designing a CAC, is to make it configurable as an integrator, an amplifier/attenuator, or, a comparator. Furthermore, the unity-gain frequency ω_o of the integrator and the gain, A, of the amplifier/attenuator should be programmable using programming devices within the CAC.

Initial effort has been directed toward choosing the appropriate approach for the proposed architecture. The CAC has a fully differential structure to attain higher CMRR and PSRR. We decided to use a current-mode approach for the following reasons:

- They are suitable for low power supply voltage.
- They offer simple primitives: the fundamental building block for current-mode circuits is the current-mirror. By using the basic block, one can implement amplifiers and integrators with programmable gains and programmable unity gain frequencies, respectively.
- They have wide-band frequency responses due to the low-impedance characteristics of the simple circuits.
- They can be manufactured using standard CMOS digital process, making it easy to integrate the FPAA with a FPGA to yield a mixed signal array.

The primitive being used is a current mirror, which can be configured to function as an integrator, an attenuator or an amplifier at its very minimum. A few other functions may also be implemented by this primitive with slight modifications. The current mode primitive is very versatile in that, it can perform the operations of signal inversion, scaling and summation. These signal processing operations are later combined to obtain more complex signal processing, including filtering and data conversion.

Several current-mode integrator circuits were considered to find the most suitable configuration. These circuits include: simple integrator, high-swing cascode integrator, and folded-cascode integrator [8,9]. To meet the requirement of a configurable current-mode integrator, the frequency response, the dynamic range, the excess phase, and the potential for adding programming elements are all important factors to be considered. Table 1 compares different integrator structures. These high frequency integrators feature good supply noise rejection and power efficiency. The comparison of these different structures led to the following conclusions.

Although the simple structure has small voltage swings, better linearity characteristics and low bias overhead requirement, the output impedance is too small which could cause serious problems when loading varies. The high-swing cascode structure has high DC gain and large output impedance, however, it requires a complex biasing circuit needed for tuning the biasing current. The bias overhead will thus be too large. The folded-cascode structure retains the advantages of the cascode circuit in terms of DC gain, output impedance, yet it is easier to be biased than the fully cascode circuit. Moreover, there is a potential to operate with a lower power supply than that needed for a fully cascoded circuit. It has also been demonstrated that the folded-cascode configuration has a good dynamic range [8,9].

Table 1. A brief comparison of integrator structures.

	DC Gain	ϕ_E	Sensitivity to Load	Configuration	Bias Overhead
Simple	low	small	high	easy	small
Cascode	high	large	low	easy	very large
Folded cascode	high	large	low	difficult	large

Fine tuning of the integrator can be done by changing the bias current. Thus a special bias circuit is needed to allow for changing the biasing current while keeping a control over the biasing voltage, so that transistors of the CAC remain in the saturation without having to change their sizes. The biasing circuit ensures constant drain to source voltages, while the biasing current is varied, to preserve the output current accuracy, when the CACs are cascaded. More details on the biasing circuit will be given in Section 3.5.

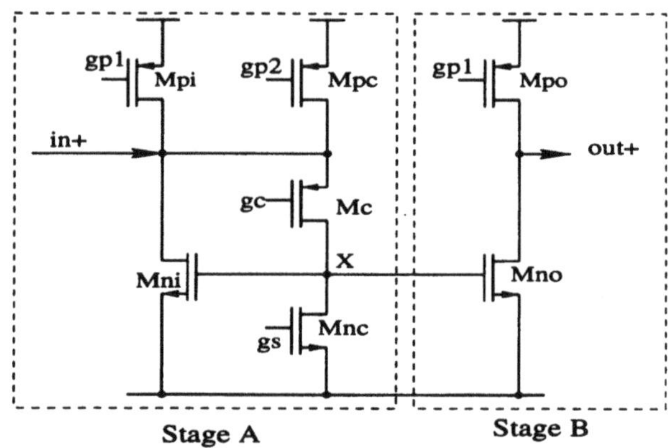

Fig. 2. Schematic of a folded-cascode current mirror.

3.2. The Proposed Topology of the CAC

Based on the previous discussion, we have chosen two circuits as the basis for the CAC: the folded cascode current mirror (Fig. 2) and the folded cascode integrator (Fig. 3). The current mirror in Fig. 2 is a well known topology, where the current injected into Mni is mirrored to Mno. The biasing current is provided by Mpi and Mpo. To reduce the input impedance of the current mirror, a common-gate transistor (Mc) is used. It reduces the input resistance of the current mirror by a factor of g_{mc}/g_{oc}. The biasing for Mc is provided by Mpc and Mnc.

The amplifier/attenuator circuit can be constructed using this current mirror and variable numbers of stage A and stage B. Considering the parasitic capacitance at the node X to be C_p, and taking a 1:1 current mirror as an example, the transfer function of this current mirror is:

$$A_d = \frac{sC_p(g_{oni} + g_{opi}) - g_{mni}g_{mc}}{s^2C_p^2 + sC_pg_{mc} + g_{mc}g_{mni}} \tag{1}$$

where g_{mni} and g_{mc} are transconductances of mni and mc respectively, and g_{oni} and g_{opi} are output conductances of mni and mpi.

Notice that the denominator is a second order

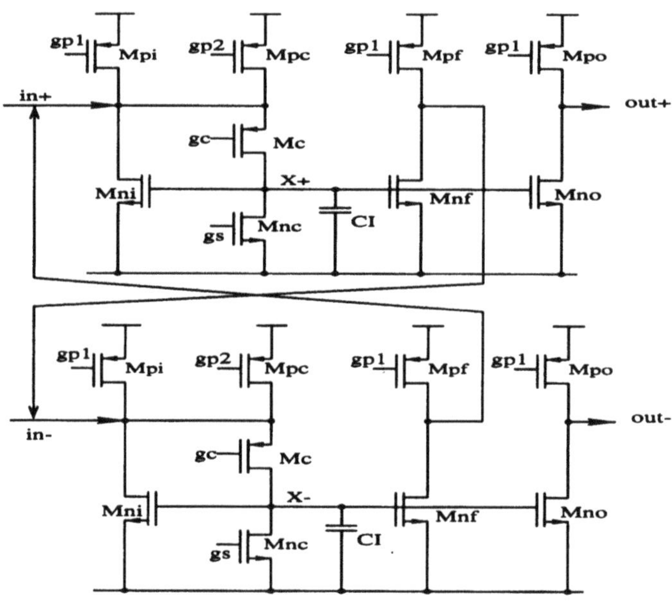

Fig. 3. Schematic of a folded-cascode current-mode integrator.

system, thus it has two poles. The location of the poles are:

$$P_{1,2} = \begin{cases} -\frac{\omega_o}{Q} \pm \frac{\omega_o}{Q}\sqrt{1-4Q^2} & \text{if } Q < \frac{1}{2} \\ -\frac{\omega_o}{Q} \pm \mathrm{j}\frac{\omega_o}{Q}\sqrt{4Q^2-1} & \text{if } Q > \frac{1}{2} \end{cases}$$

where,

$$\omega_o = \frac{\sqrt{g_{mni}g_{mc}}}{C_p}$$

and

$$Q = \sqrt{\frac{g_{mni}}{g_{mc}}}$$

We denote the current flowing through Mpi and Mni be I_1, and that through Mpc and Mnc be I_2. In order to increase ω_o, both I_1 and I_2 need to be increased. Therefore, we have to choose maximum I_1 and I_2 ($I_1 = 30\ \mu A$ and $I_2 = 30\ \mu A$).

The folded-cascode integrator, shown in Fig. 3 [8] has similarity with the current mirror (Fig. 2). Again, the input diode consists of Mni and Mc, the capacitor C_I is used for integrating the input signal and Mno provides the output current. A fully differential topology is used. The output of the middle stage Mnf/Mpf is fed back to the input of the other differential side (see Fig. 3) to cancel the common mode signal while keeping the differential gain unchanged. For more details of this integrator topology, refer to [8].

Neglecting the higher order effects, with C_I as the integrating capacitance, the differential current gain of the integrator is:

$$A_d = -\frac{g_{mni}}{g_{onc}} \frac{1 - sC_I(g_{oni} + g_{opi})/g_{mni}}{1 + sC_I/g_{onc}} \quad (2)$$

The unity-gain frequency of the integrator is:

$$\omega_o = \frac{g_{mni}}{C_I} \quad (3)$$

Note that the transistors Mni, Mnf and Mno are equally sized, and that transistors, Mpi, Mpf and Mpo are equally sized. Equation (2) indicates that the DC gain may be enhanced by decreasing g_{onc}. This can be achieved by decreasing I_2 (current through Mnc) when the CAC operates as an integrator. Equation (3) indicates that ω_o can be tuned through g_{mni} and C_I. Since it is easier to tune g_{mni}, C_I is kept constant. C_I can take one of two values, based on the desired frequency range. For example, the 100 pF capacitor enables operation down to 10 kHz and 10 pF permits operation up to about 10 MHz. Coarse and fine schemes are proposed for the adjustment of g_{mni} to tune ω_o with reasonable precision. For the coarse adjustment, g_{mni} is multiplied by an integer value β. This is done by scaling all transistors in the integrator by β. The fine tuning is accomplished by varying the DC biasing currents in small steps. The extent of fine tuning is determined by the resolution of the current generator or the biasing circuit, described in Section 3.5.

Simulation of the integrator, demonstrates the

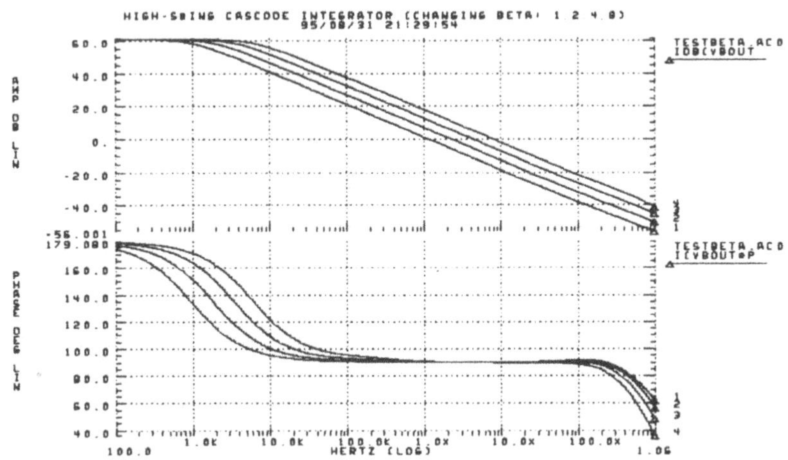

Fig. 4. Programming the integrator by changing β ($\beta = 1, 2, 4, 8$).

programmability of the unity-gain frequency as shown in Fig. 4. It illustrates the tuning of ω_o by choosing different value of β ($\beta = 1, 2, 4, 8$).

Characteristics of the CAC will be directly reflected on the performance of the circuit constructed using the CACs. One of the factors of prime concern in a CAC is the excess phase. Phase errors lead to an inaccurate quality factor of the resulting filter. While, it would be desirable to have an integrator with infinite quality factor, attempts have been made to decrease the phase errors and minimize the passband error. Feedback compensation in the CAC maintains the excess phase of the integrator within 1°. This method of compensation is very straight forward, and introduces a pole zero pair, using a series RC combination.

3.3. Implementation of the CAC

Fig. 5 shows the proposed CAC architecture. It consists of n identical slices connected in parallel at the hardwired nodes (in+/in−), (out+/out−) and the integrating node (X+/X−). The number of slices (n) can be anywhere between 1 and 10. Unscrupulously increasing the number of slices to enhance the programmable range may lead to large parasitics at the hardwired nodes. As shown in Fig. 6(a), the first slice is slightly different from the rest (Fig. 6(b)). The difference is in that, the first slice is devoid of the switches S1i. This is done to decrease the complexity of the logic circuit as described in Section 3.6. Each slice consists of five structurally similar branches. These are Mni/Mpi, Mnf/Mpf, Mno1/Mpo1, Mno2/Mpo2, and Mno3/Mpo3. All the n channel transistors in these five branches are sized equally. So also is the case with the p channel transistors. Ignoring the Mno2/Mpo2 and Mno3/Mpo3 branches, the slice shown in Fig. 6(a) and (b) has an obvious resemblance to the integrator of Fig. 3. If the Mnf/Mpf branch is omitted, the circuit is similar to the current mirror of Fig. 2. The Mno2/Mpo2 and Mno3/Mpo3 branches are added to provide additional output tabs that may be necessary, if more than one fanout is needed. The slice structure can hence be configured as a current mirror or integrator with up to 3 output tabs.

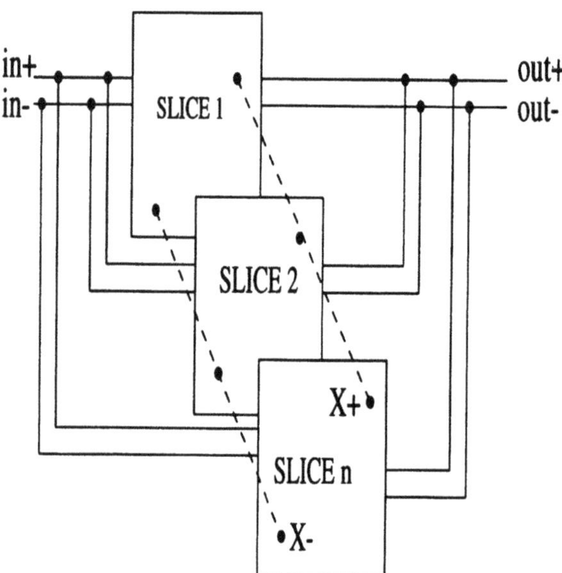

Fig. 5. The proposed topology of the CAC.

The CAC can be programmed for functionality as well as performance parameters. Programmability for both the functionality and the parameters can be achieved by activating certain branches in a stipulated number of slices. All branches are by default, deactivated. A deactivated branch has all its transistors turned off. Note that no gates are left floating. Any transistor which is to be turned off will have its gate connected to the proper rail. The process of activating a branch will be described in the next paragraph. To construct a current amplifier with a current gain "m", the Mni/Mpi, Mnc/Mc/Mpc and the Mno1/Mpo1 branches of the first cell are activated. In the remaining "m-1" cells, only the Mno1/Mpo1 branches are activated. The Mno2/Mpo2 and the Mno3/Mpo3 branches may be selected in all the slices, depending upon the number of outputs necessary. For an attenuator with "1/m" loss, the Mni/Mpi, Mnc/Mc/Mpc and the Mno1/Mpo1 branches of the first cell are the only active branches. In the other "m-1" slices only the Mni/Mpi branches are selected. As for the integrator, its unity gain frequency ω_o can be tuned through the β factor as explained in Section 3.2. This factor corresponds to the number of active slices. Also, in this case, all branches in a particular slice should be activated, except for the output tabs, Mno2/Mpo2 and Mno3/Mpo3, which are turned on only if required. Tables 2 and 3 summarize the programming states of the CAC.

This CAC structure offers a compromise between flexibility and area efficiency. Programming switches add parasitics to the CAC. Particularly, when the

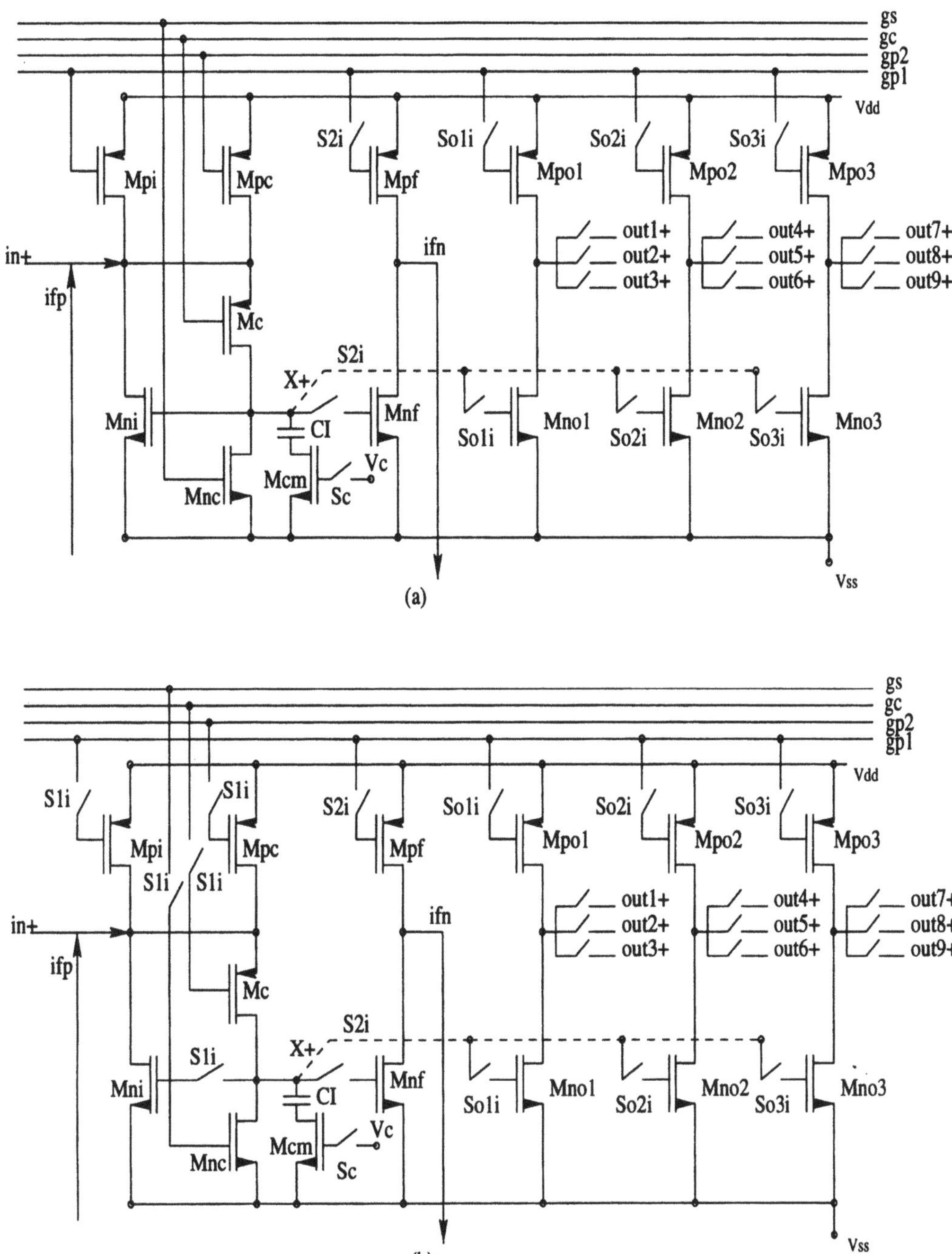

Fig. 6. (a) Circuit schematic of the first slice of the CAC, (b) Circuit schematic of the other slices of the CAC.

switch is in the signal path, it adds internal signal carrying nodes. Every switch carries with it a parasitic capacitance as well as a finite resistance and at least one pole is introduced into the system for every such node. Thus, it is preferable to put a minimum number of switches in the signal path. To achieve this, most of the signal path is hardwired. Note that the inputs (in+/in−), the outputs (out+/out−) and the integrating node (X+/X−) of all slices in a CAC are hardwired. A feature of this CAC structure is that

Table 2. Programming states of the first slice in the CAC.

Function	Stage A	Stage B	Stage C
integrator	on	on	on
amplifier	on	off	on
attenuator	on	off	on

Table 3. Programming states of other slices in the CAC.

Function	Stage A	Stage B	Stage C
integrator	on	on	on
amplifier	off	off	on
attenuator	on	off	off

most of the switches used to activate a slice or a part of it, are in the biasing path. A similar approach has been proposed in a bipolar based FPAA [5]. While the transistors Mpi, Mpf, Mpo, Mpc, Mc and Mnc can be controlled through the biasing lines, the same is not the case with the transistors, Mni, Mnf and Mno (see Fig. 6). To activate or deactivate them, a switch would be needed, either at their gate or their source. A signal path switch placed at the source will carry current and drop a finite voltage across itself, with a subsequent degradation of the frequency performance. On the other hand, if the switch is connected to the gate of the Mni, Mnf, or Mno transistors, it carries nearly zero current. Hence, the latter of the two options has been employed to activate the Mni, Mnf and Mno transistors. In essence, the proposed architecture requires fewer switches in the signal path. The analysis and simulations indicate that the impact of these signal path switches on the AC performance of the circuits is negligible in the frequency range of interest. The transistor labeled Mcm in Fig. 6, serves a dual purpose. It acts as the active resistor to introduce a zero to compensate for the phase error of the integrator and at the same time it also acts as a switch to connect the capacitance to the integrating node when the CAC is to be configured to behave as an integrator or otherwise.

The routing word controls the connection of a given CAC to its neighboring CACs. Within each CAC there are three outputs, which may be turned on or off, upon demand. Every one of these output tabs is switchable (through So1i, So2i and So3i) to get rid of the offset current, which would exist without the switch control. In the absence of these switches, if the output node is not loaded by another CAC, then the large voltage swing at this output node gets coupled to the integrating node through the gate oxide capacitance, resulting in an offset current. As shown in Fig. 6, each cell has three current outputs, with the first output connected to the out1, out2 and out3, second output to out4, out5 and out6 and the third output to out7, out8 and out9. To illustrate where the outi($i = 1, 2, \ldots, 9$) is connected, take CAC5 in Fig. 1; as an example: We define out1, out2 and out3 are inputs of upper three neighborers, as those of CAC1, CAC2 and CAC3 in Fig. 1; out4, out6 are inputs of left hand and right hand neighborers, as those of CAC4 and CAC6. Likewise, out5 is input of the cell itself, and out7, out8, out9 are inputs of the lower three neighborers, as those of CAC5, CAC7, CAC8 and CAC9 in Fig. 1. The presence of switches in the signal path, is not however detrimental to the AC performance, especially with appropriately sized switches (e.g. 24 μm/2 μm for a 2 μm CMOS process). Simulations have indicated that, up to three of them in series, present in the signal path, may be used without significant loss of frequency domain performance.

Although there is provision for widespread interconnection over the array, it is vital to evaluate the type of interconnections necessary. While, global connections provide a more general array, it is not very feasible to do so in a FPAA, since it would add a substantial amount of parasitic capacitance and resistance, which would result in the degradation of frequency response. Moreover the practicality of using local interconnections in a FPAA has been pointed out in Section 2.

3.4. *Results and Simulations of the CAC Implementation*

The proposed FPAA has been designed in a 2 μm ORBIT digital CMOS process. It has been demonstrated that the proposed FPAA can operate over almost 3 decades of frequency range (30 kHz to 10 MHz). Individual cells have been simulated. Simulations of a CAC, configured as an integrator, demonstrate the programmability of the unity-gain frequency as shown in Fig. 4. It illustrates the programming of ω_o by choosing different value of β ($\beta = 1$, 2, 4, 8). Note that while we program ω_o, the DC gain remains constant and the excess phase is less than 1°.

Table 4. Programming parameters of the CAC configured as an integrator.

Cap (pF)	Number of Slices	Comp. Tran. Size	Ibias (μA)	*fo*	Vc (V)	ϕ_E (deg)
100	1	70/2	3	39.81 K	0	− 0.04
100	1	70/2	15	74.43 K	0	− 0.37
100	1	70/2	30	95.5 K	0	− 0.27
100	5	70/2	3	172.99 K	1.5	0.54
100	5	70/2	15	360.17 K	1.5	0.63
100	5	70/2	30	469.96 K	1.5	0.77
10	1	6/2	3	248.37 K	− 0.2	− 0.29
10	1	6/2	15	521.34 K	− 0.2	0.15
10	1	6/2	30	706.27 K	− 0.2	− 0.79
10	5	6/2	3	1.04 M	1.5	− 0.81
10	5	6/2	15	2.23 M	1.2	− 0.37
10	5	6/2	30	3.13 M	1.0	0.4
10	10	6/2	3	3.13 M	− 0.45	− 0.77
10	10	6/2	15	5.9 M	0	− 0.74
10	10	6/2	30	7.15 M	0.5	− 0.07

Table 4 provides an example of the programming parameters, to cover the frequency range of interest. It also provides excess phase data for each case. It gives an estimate as to the number of slices to be activated, the biasing current required and the capacitance value, to achieve a certain unity gain frequency for an integrator. For example, to construct an integrator with a unity gain frequency of 1.04 MHz, five slices of the CAC should be activated, and a biasing current of 3 μA should be used. The capacitance value to be chosen for this example is 10 pF, which is realized using the gate oxide capacitance of an NMOS transistor. As can be inferred from Table 4, the different cases cover a wide frequency range, by using only two values of capacitors (10 pF and 100 pF). It should be noted that the compensation for the CAC is designed such that the aspect ratio of the compensation transistor, Mcm (see Fig. 6) is the same for the two extremes of frequencies covered by a particular capacitance value. Thus, only two distinct transistors serve this purpose; a 70 μm /2 μm and a 6 μm/2 μm transistor for the 100 pF and 10 pF cases, respectively. In a 2 μm CMOS technology, the area of a complete CAC (with $\beta = 10$ and two 100 pF integrating capacitors) is 2 mm^2, as estimated from the layout. For a 0.5 μm technology, however, the area of a CAC will be scaled down to 0.2 mm^2. This would allow for integrating at least 100 CACs on a single chip with an area of $0.5 \times 1\ cm^2$ taking into account the area required for metal wiring.

3.5. *Biasing Circuit*

The fine tuning of the integrator parameter (i.e ω_o) is achieved through changing the biasing current as explained in Section 3.2. The precision of the current is crucial. It is also important that the drain-to-source voltages of the transistors in the CAC remain unchanged as the biasing current varies, so that all transistors operate within the saturation region for the desired biasing current range. Both requirements are satisfied by using a special biasing circuit shown in Fig. 7. It uses OPAMPs for this purpose. The specifications of these OPAMPs are not very stringent. It is required that they have a reasonable gain and good phase margin, for maintaining stability. For this reason, the OPAMPs employed are simple differential pairs.

The precision of the current reference is guaranteed by using a precise resistance, R (could be external). The voltage across R is determined by Vdd and Vref. This reference current, (Vdd-Vref)/R is mirrored by stage B. The current gain of the mirror is determined by the number of transistors Mnxi$(i = 1, 2, \ldots, k)$ which can be switched upon demand. Note that Mpx and Mpy are replicas of the Mpi in the CAC and Mnz is a replica of Mni. Similarly, Mpc, Mc and Mnc are replicas of Mpc, Mc and Mnc respectively, in the CAC of Fig. 6(a) and (b). The biasing current injected into Mpx will generate the appropriate voltage, gp1, which will ascertain that the drain of Mpx remains at Vref, due to the action of the OPAMP. When gp1 is fed to any of the Mpi, Mpf and Mpo transistors in the CAC,

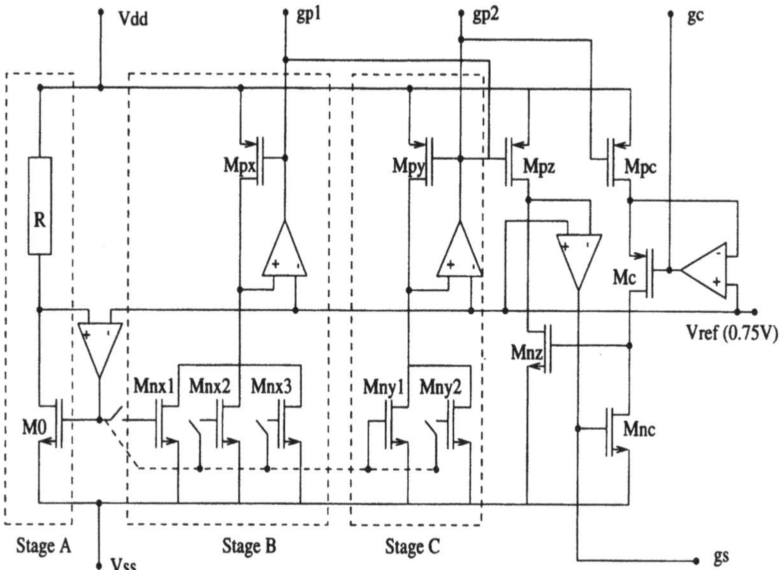

Fig. 7. Biasing circuit.

its current will be equal to the biasing current of Mpx and its drain-to-source voltage will also be close to Vdd-Vref. Similarly, gp2, gc and gs are generated in a fashion that will keep Mpc, Mc and Mnc in saturation for all possible values of the biasing current. Recall from Section 3.2 that the current I_2 through the transistor branch Mpc/Mc/Mnc of the CAC, is different for an integrator and for a current mirror. To enable this, the control is incorporated in the biasing circuit itself, so as to avoid additional switches in the CAC. Transistor Mny2 is switched to control the current through the transistor branch Mpc/Mc/Mnc. In other words the transistor Mny2 is turned on only when the CAC is to be configured as a current attenuator or amplifier. This is essential to increase the bandwidth of the attenuator or the amplifier, as shown by the analysis in Section 3.2. The function word (FW) in the programming register (PR) is responsible for identifying the value of I_2. The parameter word (PW), on the other hand can vary the biasing current, I_1 by switching transistors Mnxi.

3.6. *Programming Logic*

The functionality and the parameters of the CAC are defined through the Programming Logic (PL), shown in Fig. 8. The control circuit is fairly simple. Fig. 9 delineates the different sections of the control circuit. The control is provided by the bits in the programming register (PR).

The PR of every CAC holds 25 bits of programming information as depicted in Fig. 8. Function Word (FW) defines the functionality of the CAC and is made up of two bits, Pr0 and Pr1. The 2:3 decoder, shown in Fig. 9(a) generates three signals f_{int}, f_{amp} and f_{attn}, corresponding to each of the functions. These signals are in turn used by the other logic blocks to generate other controls. The Parameter Word (PW) begins with 9 bits (Pr2, Pr3 . . . , Pr10) for programming the number of slices to be activated in a CAC. Note that the first slice has no S1i switches as shown in Fig. 6. Instead, those nodes are hardwired to the biasing lines. Switches S2i in the first slice are controlled by the signal f_{int}, generated by the bits Pr0 and Pr1. Since the branch Mpf/Mnf provides the cross connection for gain enhancement, it is activated by the S2i switches, only when the CAC is to be configured as an integrator. The So1i, So2i and So3i switches in the first slice of the CAC are controlled by the output tab selection bits Pr16 through Pr24 which will be introduced later in this section. This topology of the first slice obviates the use of a dedicated bit for the activation of the first slice, since a minimum of one slice is always needed.

Bits Pr2 through Pr10 are responsible for activating

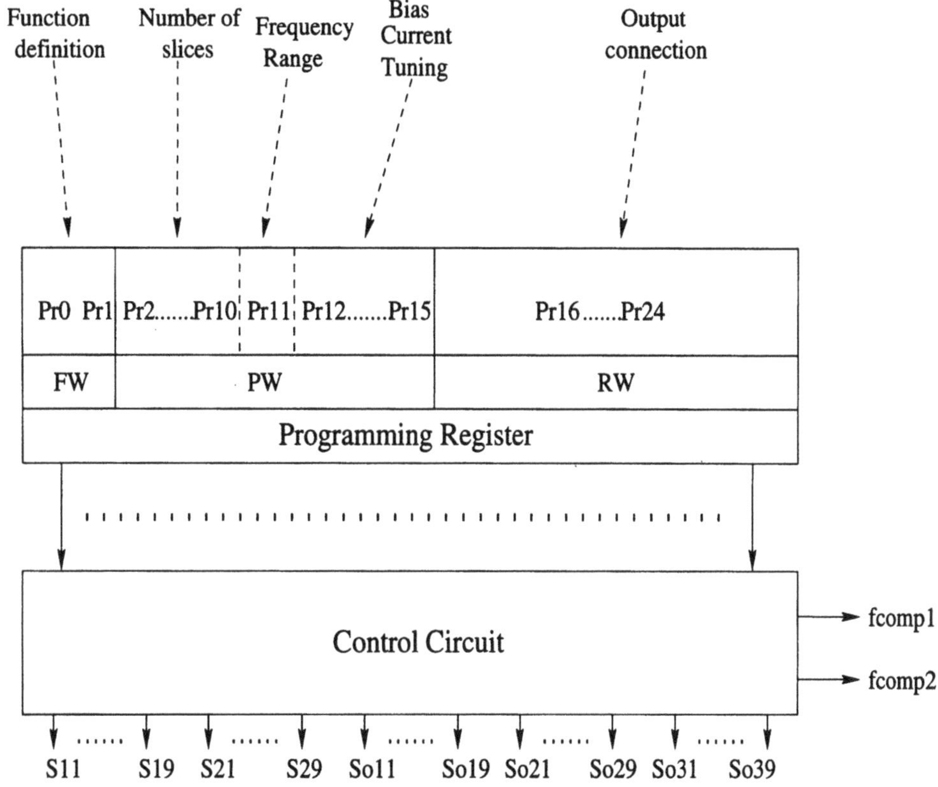

Fig. 8. The programming register.

slices 2 through 10, respectively. The signals, s1i, s2i, so1i, so2i and so3i$(i = 2, 3, \ldots, 10)$, control the switches S1i, S2i, So1i, So2i and So3i respectively, in the second through the tenth slice. For example, the switches, S1i are to remain deactivated when the CAC is configured as an amplifier. Hence $\overline{f_{amp}}$ is used as one of the inputs to the AND gate with Pri$(i = 2, \ldots, 10)$ as the other, to generate the control for the S1i switches. Similarly, f_{int} along with Pri$(i = 2, \ldots, 10)$ generates the control for the S2i switches (see Fig. 9(b)). The PW also dictates the choice of the capacitance value depending upon the desired frequency range. The signals f_{comp1} and f_{comp2} are utilized to choose one of the two available capacitors. These signals are generated by using one bit, Pr11 and f_{int} as shown in Fig. 9(c). Bits Pr12 through Pr15 are used to program the biasing current, I_1 referred to in the previous section. The number of bits used for this purpose, account for the resolution of the fine tuning scheme. These bits switch transistors Mxi in the biasing circuit shown in Fig. 7. With four bits used for programming the current, it is conceivable to have 16 levels of biasing current to choose from. $\overline{f_{int}}$ is employed to switch on transistor Mny2 in the biasing circuit (Fig. 7) to increase the current I_2 through the branch Mpc/Mc/Mnc of the CAC for increasing the bandwidth of the attenuator or amplifier as explained in Section 3.2. The PW thus amounts to 14 bits.

The Routing Word (RW) is composed of 9 bits, Pr16 through Pr24, which are used to regulate the connectivity between different CACs and for the output tab selection. Each output is connected to the inputs of three neighboring cells, as explained in Section 3.3. At most one of these three connections is the valid connection. Pr16, Pr17 and Pr18 configure the output connections for the output one, moreover, if one of them is activated, the first output is a valid output, thus ots1 (output tab selection for the first output) will be set to 1. The ots1 can be generated by

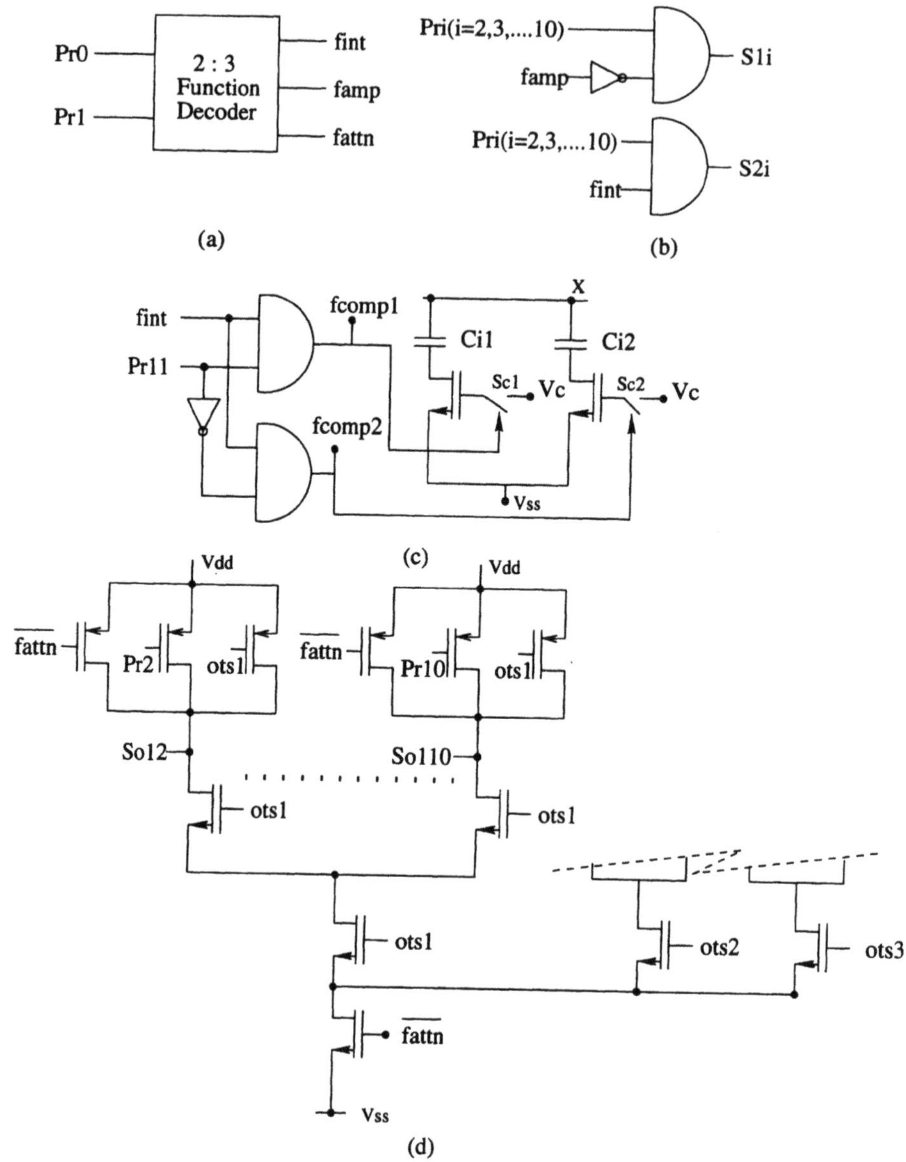

Fig. 9. Subsections of the control circuit.

the three input OR gate, with Pr16, Pr17 and Pr18 as inputs. Likewise, Pr19 through Pr21 establish the output connections for the second output and generate the signal for ots2, and Pr22 through Pr24 are responsible for the third output. The logic block exercising control over the output tabs is shown in Fig. 9(d). The circuit is analogous to a three input NAND gate with, $\overline{f_{attn}}$, Pri$(i = 2, 3, \ldots, 10)$ and otsi$(i = 1, 2, 3)$ as the inputs. Since the output tabs may be activated, only if the CAC is configured as an integrator or as an amplifier, $\overline{f_{attn}}$ is used in the logic block of Fig. 9(d), to generate signals so1i, so2i and so3i. In essence, 25 bits are required to program any given CAC and its connectivity in an array.

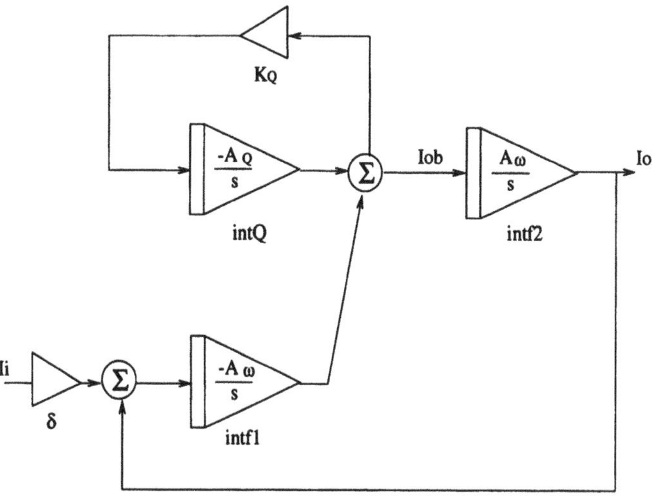

Fig. 10. A three integrator biquad structure.

4. Example of an FPAA Based Filter Implementation

The FPAA has already been employed to implement cascaded filters using biquad structures. A three integrator biquad (see Fig. 10) has been chosen for that purpose. It consists of three integrators and two attenuators. Each of those blocks is configured using a CAC. This biquad topology has the provision to independently vary the Q and the center frequency. The bandpass transfer function is given as,

$$\frac{I_o}{i_i} = \frac{-\delta A_\omega s}{s^2 + A_Q K_Q s + A_\omega^2} \tag{4}$$

$$Q = \frac{A_\omega}{A_Q K_Q} \tag{5}$$

$$\omega_o = A_\omega \tag{6}$$

A_Q and A_ω are the unity gain frequencies of the corresponding integrators. δ is the loss of the input attenuator, and K_Q is the loss of the Q attenuator. Since the unity gain frequency of the integrator is programmed by changing the number of slices and by varying the biasing current, equation (6), indicates that the center frequency of the filter can be tuned in a similar fashion. The Q of the bandpass filter thus obtained is determined by equation (5). The unity gain frequency, A_ω is designed to be equal to A_Q, so that the Q is governed solely by the loss of the Q attenuator, K_Q. Different combinations of the above mentioned parameters may be applied to realize the desired specifications.

Fig. 11 shows the mapping of the biquad onto the proposed FPAA architecture shown in Fig. 1. Six cells (CAC1 to CAC6) have been dedicated for this biquad implementation. The connectivity is established through the routing words. For example, out of the nine possible connections of the outputs of CAC5 to its neighborers, only the connections to CAC1 and CAC6 are activated, as indicated by the thick lines. The function and parameters for each cell are listed in Table 5. The functions are determined through FW (Function Word), as explained in section 4.6, the bits Pr0 and Pr1 of CAC1, CAC3 and CAC4 are set to 00 to configure them as attenuator, and similarly for CAC2, CAC5 and CAC6, these bits are set to 01 to configure them as integrator. The parameters are programmed by the PW (Parameter Word) which consists of the value of β, the range of frequency and the biasing current. The value of β for each cell depends on the applications. However, notice that CAC1, CAC3 and CAC4 have same β value which is determined by the quality factor of the biquad. CAC5 and CAC6 have similar β values for programming the biquad center frequency. Bit Pr11 is the same for all cells because they are all operating in the same frequency range. The biasing for each cell is provided by the biasing circuit. From Table 5, we can see that CAC1, CAC3 and CAC4 have the same biasing current (Ib_{att}) and can therefore share the same biasing circuit, while CAC2, CAC5 and CAC6 which use biasing current

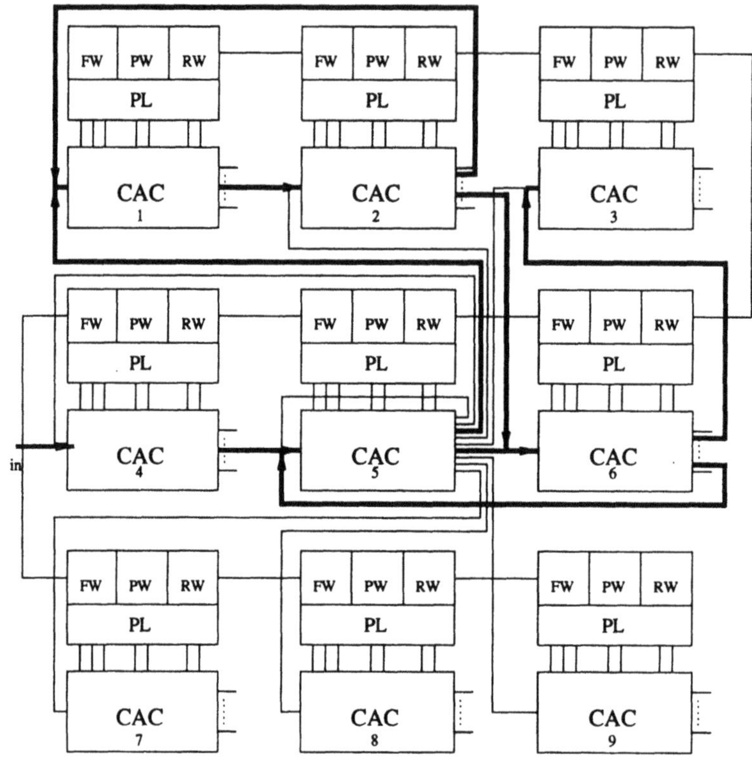

Fig. 11. Implementation of a biquad with FPAA.

Ib_{int}, share another biasing circuit. The activation of the output connections for each cell is configured through RW (Routing Word). As an example, to implement the biquad as shown in Fig. 11, Pr16, Pr21 of CAC5 are set to 1 to establish a connection between the output of CAC5 and the input of CAC1 and CAC6, while the other routing bits are not valid (have value of 0). Table 5 shows the RW for each cell.

The simulation of the biquad shows that ω_o can be programmed as shown in Fig. 12 and the programming of Q has also been verified (see Fig. 13). The programmability is achieved by both coarse adjustment (select the number of slices) and fine adjustment (vary the biasing current), on every CAC being utilized, while the compensation remains fixed.

5. Conclusion

A new architecture for a FPAA has been proposed. The architecture is unlike the conventional FPGA architecture, and eliminates the need for a switching and connection module. The architecture may be considered as a political compromise between flexibility and the number of programming elements in the signal path. As a result a unique feature of the

Table 5. Function and parameters of each cell for biquad implementation.

Cell	Function	β	Biasing	RW
CAC1	attenuator (K_Q)	β_{K_Q}	Ib_{att}	000001000
CAC2	integrator (intQ)	β_{A_Q}	Ib_{int}	000100001
CAC3	load	β_{K_Q}	Ib_{att}	000000000
CAC4	attenuator (δ)	β_{K_Q}	Ib_{att}	000001000
CAC5	integrator (intf1)	β_{A_ω}	Ib_{int}	100001000
CAC6	integrator (intf2)	β_{A_ω}	Ib_{int}	010100000

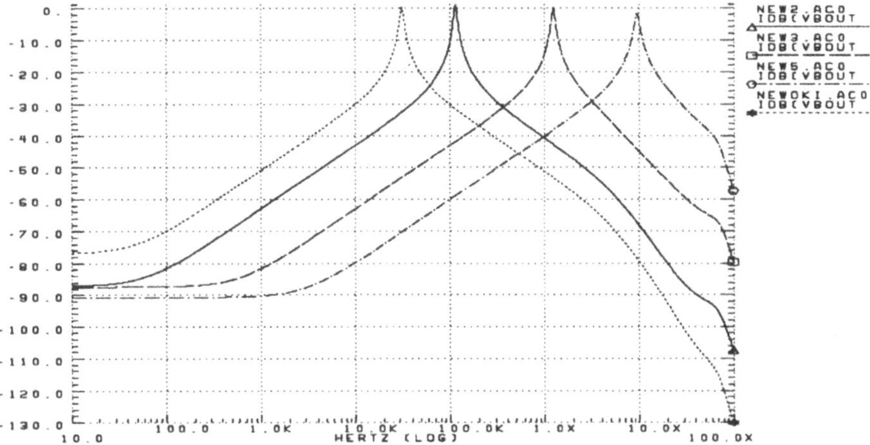

Fig. 12. Programming ω_o of a biquad structure.

FPAA is that, most of the switches are placed in the biasing path, yielding better frequency response. The FPAA is based on a simple current mode primitive. We have demonstrated the simplicity of the configuration and programming methodology. The FPAA has also been tested for feasibility by constructing programmable filters. The attributes of these filters were successfully changed and adjusted by altering the CAC parameters.

The complete programming of one CAC requires 25 bits. This scheme extends a good amount of flexibility to the FPAA. However, the overhead for programming can be reduced by a small decrease in the extent of flexibility. One of the options would be to divide the frequency range into two and thus create two categories of FPAAs but with the same primitives. Alternately, the number of connections possible between any individual CAC and its neighbors may be decreased, or the resolution of fine tuning can be reduced to diminish the number of programming bits. As an epitome, the FPAA presented in this paper may be modified with minimum inconvenience to satiate an optimum trade off between complexity and flexibility.

An extension of the presented work, would include augmenting the flexibility of the CAC to be able to

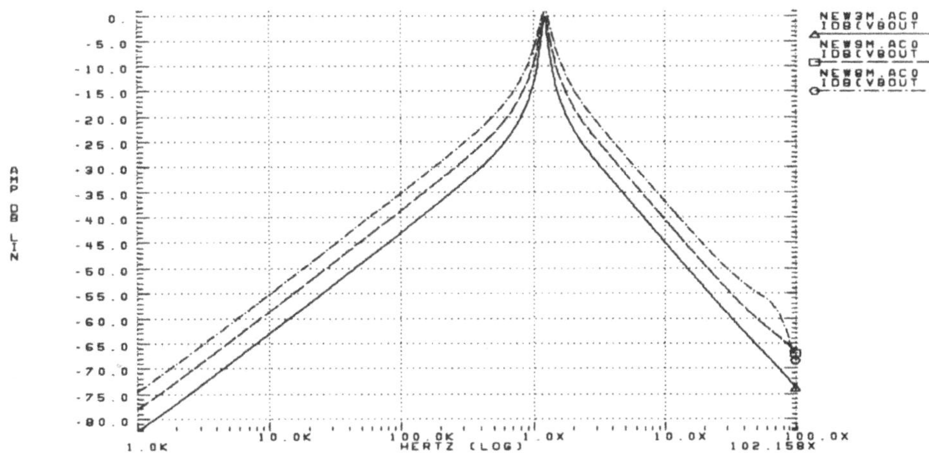

Fig. 13. Programming Q of a biquad structure.

configure it as a comparator so as to implement pipelined A/D converters using the FPAA. A CAD tool will also be developed to determine the configuration parameters like the functionality of the CAC, the number of active slices and the biasing current, given a certain set of specifications. There exists a large sensitivity of the filter cut-off frequency and quality factor, Q, to process variations and temperature variations. This in turn, is due to the dependence of the integrator unity-gain frequency on the absolute value of the transconductances and the capacitances. In addition, to these effects, the active filters are very sensitive to the integrator phase shift and its finite DC gain [10]. It is therefore, essential to design an on-chip tuning circuit for the cut-off frequency and the quality factor, Q, to overcome this problem. The focus in doing so will be on implementing a tuning circuit with the available CAC blocks thus widening the utility of the FPAA.

Notes

1. This work has been funded by a RIA grant from the National Science Foundation (Grant No. MIP-9410413).

References

1. E. K. F. Lee and P. G. Gulak, "Field Programmable Analogue Array Based on MOSFET Transconductances." *Electronics Letters*, pp. 28–29, January 1992.
2. H. Kutuk and S. Kang, "A Field-Programmable Analog Array (FPAA) Using Switched- Capacitor Techniques." *Proc. of International Symposium on Circuits and Systems* 4, pp. 41–44, May 1996.
3. E. K. F. Lee and P. G. Gulak, "A transconductor-Based Field-Programmable Analog Array." *1995 IEEE International Solid-State Circuits Conference* 38, pp. 198–199, February 1995.
4. E. K. F. Lee and P. G. Gulak, "A CMOS Field- Programmable Analog Array." *IEEE J.S.S.C.* 26(12), pp. 1860–1867, December 1991.
5. E. Pierzchala, M. A. Perkowski, P. Van Halen, and R. Schaumann, "Current-Mode Amplifier/Integrator for a Field-Programmable Analog Array." *1995 IEEE International Solid-State Circuits Conference* 38, pp. 196–197, February 1995.
6. E. Pierzchala and M. A. Perkowski, "High Speed Field-Programmable Analog Array Architecture Design." *FPGA'94*, Berkeley, Calif., February 13–15, 1994.
7. E. Pierzchala, M. A. Perkowski, and S. Grygiel, "A Field-Programmable Analog Array for Continuous, Fuzzy, and Multi-Valued Logic Applications." *ISMVL'94*, pp. 148–155, Boston, Mass., May 25–27, 1994.
8. R. Zele and D. Allstot, "Low-Power CMOS Continuous-Time Filters." *IEEE J.S.S.C.* 31(2), pp. 157–168, February 1996.
9. S. L. Smith and E. Sánchez-Sinencio, "Low Voltage Integrators for High-Frequency CMOS Filters Using Current Mode Techniques." *IEEE Trans. on Circuits and Systems II* 43(1), pp. 39–48, January 1996.
10. H. Khorramabadi and P.R. Gray, "High-Frequency CMOS Continuous-Time Filters." *IEEE J.S.S.C.* SC-19(6), pp. 939–948, Dec. 1984.

Sherif Embabi (s'87-M'91) received his B.Sc. and M.Sc. degrees in electronics and communications from Cairo University, Egypt, in 1983 and 1986, respectively, and the Ph.D. degree in electrical engineering from the University of Waterloo, Canada, in 1991.

In 1991, Dr. Embabi joined Texas A&M University, as an Assistant Professor of Electrical Engineering. He is currently on sabbatical leave with Texas Instruments in Dallas, Texas.

Dr. Embabi's research interest is in the area of VLSI implementations of mixed signal systems. He has worked on BiCMOS digital circuit design. Currently his interests are in the area of low voltage analog circuit design, field programmable analog arrays and RF circuits.

Dr. Embabi is a co-author of Digital BiCMOS Integrated Circuit Design, Kluwer Academic Publishers, 1993. He is the recipient of an NSF Research Initiation Award in 1994. He is currently serving as an Associate Editor for the IEEE Transactions on Circuits and Systems (Part II).

Xiaohong Quan received the B.S.E.E. and M.S.E.E. from Jiaotong University, Xi'an, China, in

1989 and 1992, respectively, and the M.S. in Physics from Texas A&M University, College Station, Texas, in 1994. She is currently a Ph.D. candidate in the Department of Electrical Engineering in Texas A&M University.

From 1989 to 1992 she was a Teaching Assistant in Electrical Engineering Department in Xi'an Jiaotong University. During the same period she participated actively in the development of Scanning Laser Acoustic Microscope. From 1992 to 1994, she was a Research Assistant in Physics in Texas A&M University, doing research in remote sensing of sound velocity and temperature of ocean using Brillouin Lidar technique. Since spring 1995, she has been a research assistant in Microelectronics group in Texas A&M University, developing Field-Programmable Analog Array for signal processing applications. The targeted applications include continuous-time filters and data converters. In summer 1996, she was doing internship in Silicon Systems—A Texas Instrument company, with its read channel department. Her present field of interest is the design of integrated circuits for telecommunications and mass storage.

Nobuo Oki received the B.S. degree from the Escola de Engenharia de Sao Çarlos—Universidade de Sao Paulo in 1977 and M.S. degree from Centro de Ciencias e Tecnologia—Universidade Federal da Paraiba in 1980. He received his Ph.D. degree in 1989 from Faculdade de Engenharia Eletrica at Universidade Estadual de Campinas, Sao Paulo, Brazil.

From February 1981 he has been with the Electrical Engineering Department at Universidade Estadual Paulista—UNESP, Ilha Solteira, Sao Paulo, Brazil, where he is currently an Assistant Professor. During the years of 1994 to 1995 he joined the Microelectronics Analog Group at the Texas A&M University as a Visiting Scholar. His main research interests are analog integrated circuits and neural networks implementation. He is a member of IEEE.

Ashish Manjrekar received the B.E. degree in Instrumentation Engineering in 1993, from the University of Bombay, India and the M.S. degree in Electrical Engineering in 1997 from Texas A&M University, College Station, TX. Since August 1996, he has been with the Mixed Signal Products group, Texas Instruments, Dallas. At Texas A&M University he was involved in the design of current mode filters and amplifiers, besides the development of a Field Programmable Analog Array. His areas of interest include CMOS current mode circuit design and mixed mode integration.

Edgar Sánchez-Sinencio received the M.S.E.E. degree from Stanford University, Stanford, California, and the Ph.D. degree from the University of Illinois at Champaign-Urbana, in 1970 and 1973, respectively. He did an industrial postdoctoral with Nippon Electric Company, Kawasaki, Japan in 1973–174.

Currently, he is with Texas A&M University, Department of Electrical Engineering, as a Professor. he is the co-author of Switched-Capacitor Circuits (Van Nostrand-Reinhold, 1984) and co-editor of Artificial Neural Networks: paradigms, Applications, and Hardware Implementations (IEEE

Press, 1992). He has been the Guest Editor or Co-Editor of three special issues on Neural Network Hardware (IEEE Trans. Neural Networks, March 91, May 1992, May 1993) and one special issue on Low Voltage Low Power Analog and Mixed-Signal Circuits and Systems (IEEE Trans. Circuits and Systems I, November 1995). His research interests are in the area of solid-state signal processing circuits, including BiCMOS, CMOS Neural Networks, Fuzzy and RF-circuit implementations. He was the IEEE/CAS Technical Committee Chairman on Analog Signal Processing (1994–1995). He has been Associate Editor for different IEEE magazines and transactions since 1982 until now. He was the IEEE Video Editor for the IEEE Trans. on Neural Networks. He was also the IEEE Neural Network Council Fellow Committee Chairman 1994 and 1995. He was a member of the IEEE CAS Board of Governors (1990–1992). He is the 1993–1994 IEEE Circuits and Systems Vice President-Publications and a member of the IEEE Press Editorial Board. In 1995, he received an Honoris Causa Doctorate from the National Institute for Astrophysics, Optics and Electronics, Puebla, Mexico.

In 1992 he was elected as a Fellow of the IEEE for contributions to monolithicanalog filter design.

Analog Integrated Circuits and Signal Processing, 17, 143–156 (1998)

A High-Frequency Field-Programmable Analog Array (FPAA) Part 1: Design

EDMUND PIERZCHALA AND MAREK A. PERKOWSKI

Department of Electrical Engineering, Portland State University, Portland, OR 97207-0751
edmundp@ee.pdx.edu, mperkows@ee.pdx.edu

Received August 2, 1996; Accepted November 10, 1997

Abstract. The design of a high-frequency field-programmable analog array (FPAA) is presented. The FPAA is based on a regular pattern of cells interconnected locally for high frequency performance. No switches of any kind are used in the signal path of a cell: programming of the functions, parameters, and interconnections is achieved solely by modifying cells' bias conditions digitally. Limited global signal interconnections are also available for those application circuits which cannot be mapped onto locally-only interconnected structure. Key circuits of the FPAA have been fabricated in a CPI transistor-array bipolar technology.

Key Words: programmable circuit, field-programmable analog array (FPAA), current-mode circuit, analog signal processing

1. Introduction

Field-Programmable Gate Arrays (FPGAs) have found many applications since they were proposed about a decade ago. FPGAs dramatically shorten design time and allow instantaneous modifications and corrections. Their applications range from simple "glue logic" functions to complex, dynamically reconfigurable systems.

The success of FPGAs is undoubtedly one of motivating factors in the FPAA research. With the current trend which favors digital techniques, analog circuits seem to be left to perform interface functions (such as A/D, D/A converters, anti-alias and smoothing filters) or work where digital circuits did not yet achieve satisfiable performance (e.g. high-frequency applications). It seems that analog circuits, as more capricious and harder to design, should yield to digital ones. In fact, the picture is not so simple, and the "digital revolution" relies heavily on progress in analog circuits in each of its victories [21]. The design complexity of analog front-end and back-end circuits may exceed the complexity of the digital signal processing circuit they work with. As an example one can consider processing of video signals, or any other signals of sufficiently high bandwidth. In such applications, the sampling frequency is often chosen to be close to the Nyquist frequency, which poses stringent requirements on the anti-aliasing filter design, leading to high-order filters. Moreover, if high linearity and low noise are desired, one must not only assure high quality of data converters, but also the analog front and back ends. All in all, the analog part of the entire system may easily become as or even more complex than the digital one. At that point one may ask if the implementation of the entire signal processing channel as an analog one would not be a better choice [19].

Analog circuits can perform important signal-processing functions, such as multiplication and integration, faster, using less power, and on less silicon real estate than their digital counterparts. For these and other reasons, mentioned earlier, analog circuits are rather unlikely to be eliminated entirely from the electronic design, nor to be reduced to some simple, residual form in a predominantly digital design world. Therefore, it is of utmost importance to ease the analog design process.

One of the reasons analog design is so much more complex than digital, is that the number of design options and trade-offs it involves is much larger than in the digital realm. Also, analog designers have significantly less freedom in ignoring low-level circuit interaction of high-level blocks in a hierarchical design. A carefully designed multi-function analog circuit such as an FPAA can successfully address these issues, delivering the full potential of analog circuits to a designer, who may or may not be an analog expert.

A number of analog programmable circuits have been reported in the literature. Due to the use of switched capacitor techniques [2], subthreshold MOS operation [10,11], or extensive use of global signal interconnections [12,13], these devices have limited bandwidth and are generally not suitable for high-frequency operation. An extensive review of prior work in FPAAs is presented in [1].

This paper presents results of research aimed at developing programmable analog circuits for high-frequency applications, the first attempt to build such circuits reported in the literature. Preliminary results were presented in [16].

Given that the semiconductor technologies advance rapidly, "high frequency" in this paper is not defined in terms of numbers, but rather as an attribute of an electronic circuit to operate at, or near to (e.g. within one order of magnitude), the maximum signal frequency supported by a given technology. Using this convention, a 1.2 μm CMOS circuit operating at 30 MHz will classify as a "high-frequency" one, while a 100 MHz circuit realized in a 27 GHz bipolar process would rather be considered a low-frequency one.

The paper is organized as follows. Section 2 addresses the question of architecture (i.e. pattern of interconnections) most suitable for high-frequency operation. Section 3 describes the design of the analog circuits of a single cell of the FPAA. Section 4 presents the design of the digital control circuit of a single cell. Finally, Section 5 contains conclusions.

2. Architecture

2.1. *Background*

It is well known that the high-frequency performance of an analog circuit built in any specific technology depends on the particular circuit techniques used, and the layout. In this section the focus is on these aspects of the design which determine the geometric properties of the programmable device and lead to a particular layout.

An FPAA consists of individual signal-processing blocks (cells) and signal interconnections between them. Layout techniques used for traditional (i.e. non-programmable) circuits are insufficient for programmable devices, as the latter must have redundant interconnections for the sake of programmability. Likewise, most layout (or architectural) techniques developed for digital programmable devices (such as FPGAs or PLDs) are unsuitable for high-frequency FPAAs, because analog circuits cannot tolerate delays, phase errors, and cross talk between signals that digital circuits can.

An architecture, or topology, of a programmable device, comprises two elements: the design and the resulting functionality of a single cell, and the pattern of interconnections between cells. What is considered a single cell, is to some degree an arbitrary decision. A cell in one programmable device might be considered a collection of cells, or a "macro cell," in another, giving rise to a hierarchical architecture. It is convenient to think of a single cell as a unit capable of performing some elementary signal processing functions, such as integration or amplification, which can be combined according to the topology of the programmable device in order to realize desired circuit or system function. For instance, if one is interested in realizing linear filters, a cell capable of implementing an arbitrary second-order function seems like a reasonable choice.

It is assumed in this paper that the cell is autonomous, i.e. it is capable of operation on its own, without the presence of any other cells. Thus a single transistor could not be considered a cell, even though by a suitable interconnection of transistors one can realize a host of circuits.

The signal interconnections must connect these cells that need to be connected, while at the same time they must provide adequate isolation between those cells that need to remain disconnected.

A *(fully) programmable* circuit or device is one that allows changing of its configuration (pattern of interconnections between cells) as well as functions and parameters of individual cells. A *tunable* circuit is one that allows programming of parameters only. Fully programmable circuits provide more flexibility in programming than tunable ones. For instance, a tunable filter allows changing certain of its parameters, such as cut-off frequency or quality factor, whereas a fully programmable one provides the same and the means of implementing different passband configurations (band-pass, low-pass, etc.), different orders, and different approximations (e.g. Chebyshev, elliptic, etc.).

2.2. *High-Frequency Architectures*

There are two architecture schemes diametrically opposed to each other. One is based on providing programmable connections between every pair of cells in the circuit. This approach favors flexibility, but also leads to excessively long signal interconnections, which introduce phase errors and cross talk problems detrimental to the circuit operation at high frequencies.

The second scheme is based on restricting the interconnection pattern in favor of better high-frequency performance. This paper reports results of research based on the latter approach.

Intuitively, restricting the pattern of interconnections should decrease the flexibility of the programmable device, measured as the number of different circuit topologies that can be implemented. It turns out however, that this intuition is not necessarily correct. A number of important classes of circuits can be implemented in an FPAA of carefully restricted topology. It is possible because most "real-world" circuits have restricted connectivity between components; very rarely is it necessary to connect most (or all) components with most (or all) other components.

2.3. *Locally-Connected vs. Globally-Connected Topology*

Let us consider a simple pattern of interconnections shown in Fig. 1a.

Each cell (represented by a dot in the figure) can receive output signals from the four nearest neighbors, and can send its own output signal to the same neighbors. Given adequate functionality of each cell, this restricted topology allows implementation of various important classes of circuits [17], such as ladder and cascade linear filters, rank filters, modulators, demodulators, PLLs, automatic gain control (AGC).

Although a wide variety of applications can be realized in this locally-only interconnected architecture, some circuits require global connections. An interconnection pattern shown in Fig. 1b, superimposed on that of Fig. 1a but shown separately for clarity, further enables implementation of other circuits, such as matrix operations circuits, equation solvers, programming problem solvers, multi-valued logic and fuzzy logic circuits [17].

2.4. *The Cell Functions and the Control Block*

This section presents the second element comprising an architecture, namely the functionality of an individual cell. Circuit aspects of the cell design are discussed in Section 3.

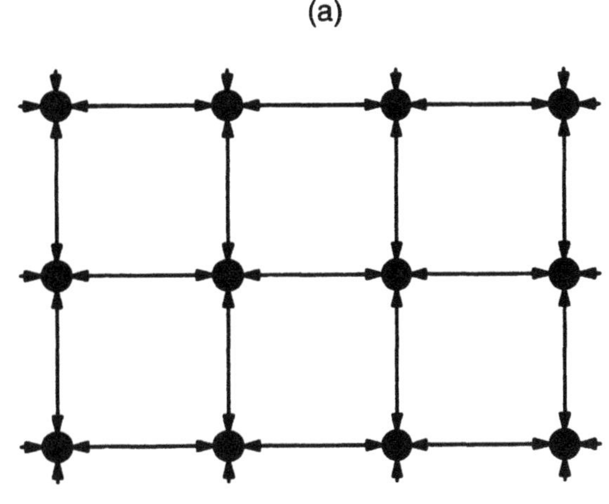

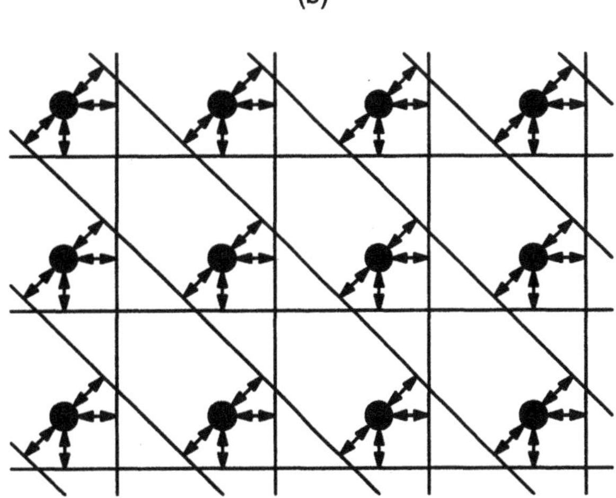

Fig. 1. Signal interconnections of the FPAA: (a) local, (b) global.

In the presented FPAA all cells are identical, but cells of different functionality could be used as well. Fig. 2 shows a functional block diagram of an individual cell. The functions and parameters of the cell are determined by the control block, presented in Section 4.

The cell works in one of the two modes: *passive-control mode* and *active-control mode*. In the passive-control mode only the analog blocks of the cell perform signal processing functions. The control circuit determines the parameters and configuration of the analog blocks of the cell, but is otherwise not involved in the signal processing. In the active-control mode, the control circuitry additionally takes part is some signal processing functions. Two important non-linear circuits are implemented in the active-control mode (see Section 4): *min/max-follower* and *controlled waveform generator (VCO, voltage-controlled oscillator)*.

2.4.1. Passive-Control Mode. As shown in Fig. 2, analog input signals are connected to two summers. When at least one weight w_i is non-zero, a summer implements weighted sum (1).

$$y(t) = k_s \cdot \frac{\sum_i s_i \cdot en_i \cdot w_i \cdot x_i(t)}{\sum_i w_i} \tag{1}$$

When all the weights w_i are zero, the output $y(t)$ is also zero. The summers' weights w_i are positive or zero, and are programmed independently, i.e. each weight w_i for one summer can be different from any other weight w_j for the same or the other summer. Each signal can be summed with positive or negative sign, s_i. Enable bits, en_i, allow connecting and

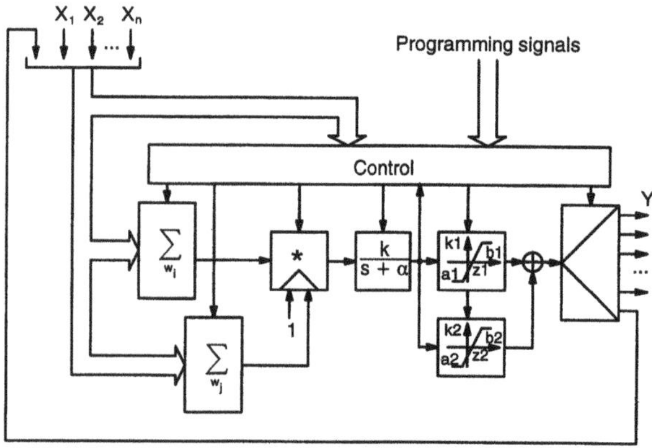

Fig. 2. Functional block diagram of the programmable cell.

disconnecting a given signal from the summer input, which is a means of programming the interconnection pattern between the cells. Signs and enable bits are programmed independently. The denominator of (1) provides scaling of the output signal dependent on the combined weights. Such scaling is necessary to ensure proper dynamic range of the output signal. The overall gain of each summer is determined by its respective k_s.

The output signals of the two summers are connected to the multiplier (2).

$$y(t) = x_1(t) \cdot x_2(t) \tag{2}$$

The multiplier also performs important signal processing functions, such as phase detection, balanced modulation and demodulation [7]. If no multiplication is needed, a constant signal, symbolically represented as "1" is connected to the second input, instead of the second summer's output.

The multiplier output is connected to the amplifier/integrator, which performs one of the three functions: amplifier, lossless integrator, or lossy integrator (3–5).

$$y(s) = k_i x(s) \tag{3}$$

$$y(s) = \frac{k_i}{s} x(s) \tag{4}$$

$$y(s) = \frac{k_i}{s + \alpha} x(s) \tag{5}$$

The output signal of the amplifier/integrator is connected to a pair of limiting (clipping) blocks, each of which realizes the basic DC transfer function represented by (6), also illustrated in Fig. 3a.

$$y = \begin{cases} -a & \text{if } x < -a \\ x & \text{if } -a \le x \le a \\ a & \text{if } x \ge a \end{cases} \tag{6}$$

This basic clipping characteristic of each block can be electronically shifted along the vertical and horizontal axis, and the slope of the linear part can be changed, as shown in Fig. 3b. By combining (adding) two clipping characteristics one can obtain a variety of nonlinear DC transfer functions. Some examples are shown in Fig. 3c–h. Such important functions as **abs** (full-wave rectifier, Fig. 3e), **sign** (Fig. 3g, shifted along the horizontal axis) and "fuzzy-membership" (Fig. 3c, d) are easily implemented.

Output signals from the clipping blocks are added together and mirrored for distribution to the neigh-

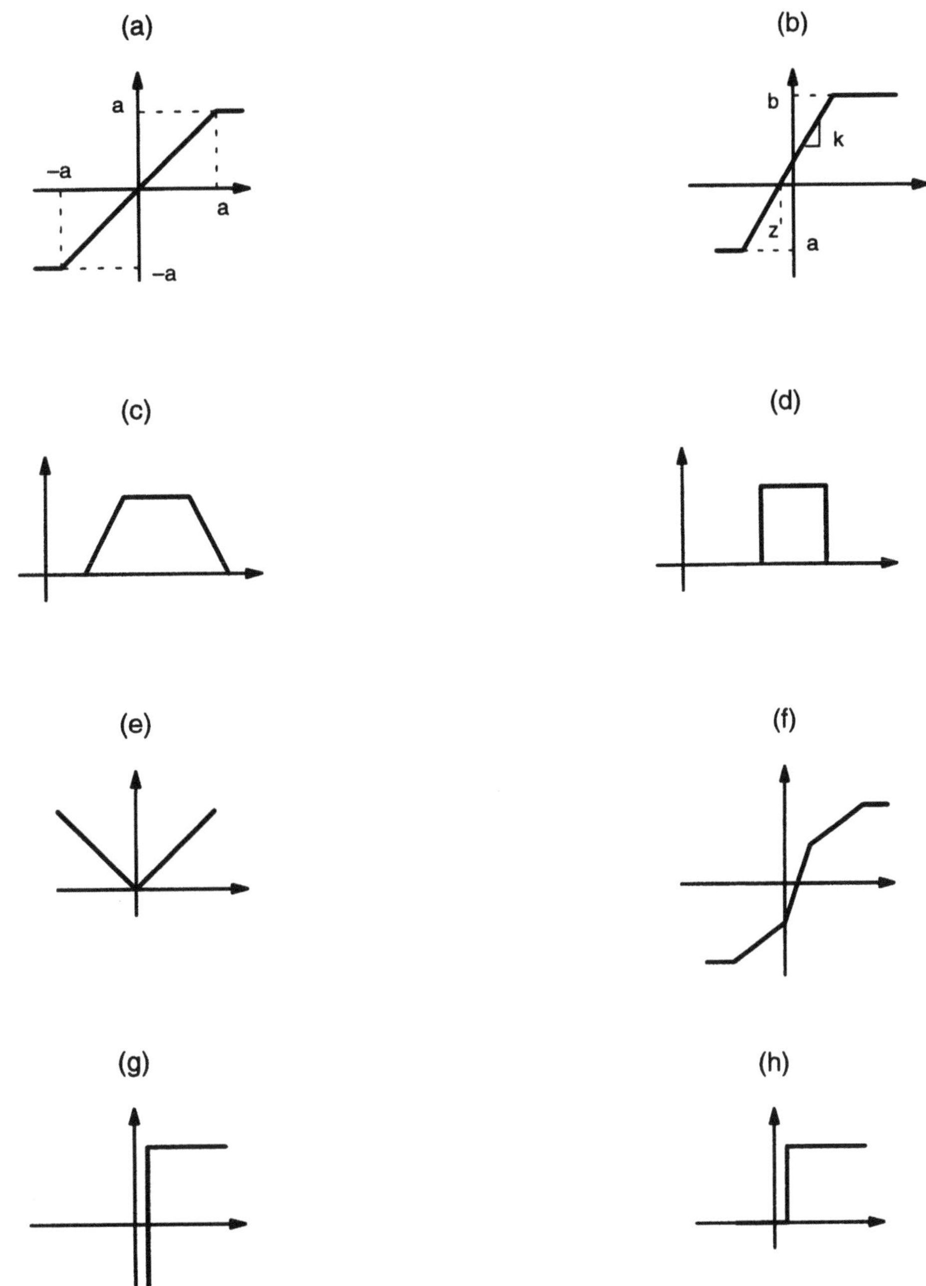

Fig. 3. Selected DC transfer characteristics of a single cell.

boring cells and global signal lines (the signals are in current mode).

There is also a "feedback" connection inside the cell, which makes the cell output signal available at the input. Some applications of the FPAA require such a connection (see the rank filter example [17]).

Table 1 summarizes the most important functions realized by a single cell, including ones in the active-control-mode.

2.4.2. Active-Control Mode. In the active-control mode, the analog processing part of the cell and the

Table 1. Selected functions of a single cell.

1. $y = k \cdot \frac{\sum_i w_i x_i}{\sum_i w_i} \cdot \frac{\sum_j w_j x_j}{\sum_j w_j}$
2. $y = k \cdot \frac{\sum_i w_i x_i}{\sum_i w_i}$
3. $y = kx_i x_j$
4. $y = kx_i^2$
5. $y = k \cdot \min(x_1, \ldots, x_n)$
6. $y = k \cdot \max(x_1, \ldots, x_n)$
7. $y = k \cdot y_{1-6} \cdot \frac{1}{s+a}$
8. $y = a\ sign(y_{1-7})$
9. $y = b\ U(y_{1-7})$, U is the step function
10. $y = k\lvert y_{1-7} \rvert$
11. $y = x_i$ (identity)

control block form a feedback system which operates in a way similar to that of a data-path-and-control arrangement found in digital systems. A very complex scheme of this kind would be difficult to implement and it might be slow. In the present arrangement each of the input signals (and the output signal) can be compared against the output of the amplifier/integrator in order to control the weights and signs of the first summer. The details of the control block implementation are presented in Section 4.

3. Analog Building Blocks

Fig. 4 demonstrates the basic analog building block of the cell [3,5,7,8]. In its simplest form the circuit contains only transistors Q_1–Q_4 and the tail current source I_B^+. Current sources I_A represent the circuit's

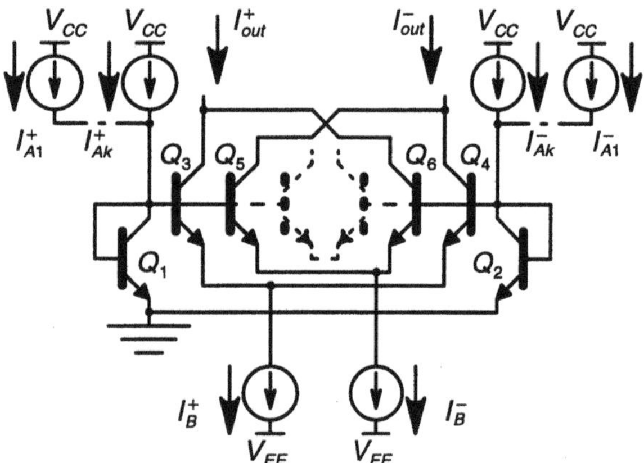

Fig. 4. Basic analog building block of the cell.

input signals. The circuit is fully differential, i.e. both input and output signals are represented by differences of currents in two wires. The sum of currents $I_A^+ = I_A(1+x)$ is the positive "half" of the input signal, and $I_A^- = I_A(1-x)$, is its negative "half." The input signal is then $I_A^+ - I_A^- = I_A(1+x) - I_A(1-x) = 2I_A x$; x is called *modulation index*. Likewise, the output signal is the difference $I_{out}^+ - I_{out}^- = I_B(1+y) - I_B(1-y) = 2I_B y$. Current gain is determined by the ratio I_B/I_A and in practice can be tuned over several decades from a fraction of unity to about 10.[1] Since there are very little voltage swings (only several hundred mV in the entire linear range of operation), the circuit has very high gain-bandwidth product, close to the f_T of the transistors [3]. In the 8 GHz bipolar process used for the implementation of the core of the cell [16] the simulated gain-bandwidth product of this circuit exceeds 6 GHz.[2]

The DC transfer characteristic of the circuit, shown in Fig. 3a for a gain of 1, exhibits sharp overload points and excellent linearity within entire linear range. The width of the linear part of the characteristic and its slope are determined by the bias currents I_A and I_B. By adding (subtracting) currents on the input and on the output of the circuit (by additional programmed current sources) one can change the location of the zero of the characteristic, as well as the two clipping (saturation) levels (Fig. 3b).

This circuit has many variations; all the remaining analog blocks of the cell either contain one of those variations directly, or are related to one. For instance, including transistors Q_5 and Q_6 (Fig. 4) allows inverting the signal (negative weight). If another pair of inputs is connected in place of the tail current sources I_B^+ and I_B^-, the circuit becomes a Gilbert multiplier core [4]. More transistor pairs can be added (dashed line) to obtain several outputs, such as it is required to implement a differential current mirror. Each output can be independently tuned by means of changing its tail current.

3.1. Summer

Fig. 5 shows the schematic of a summer with independent tuning of input weights. Additional summation (without independent tuning) can be realized by connecting several signals to each input.

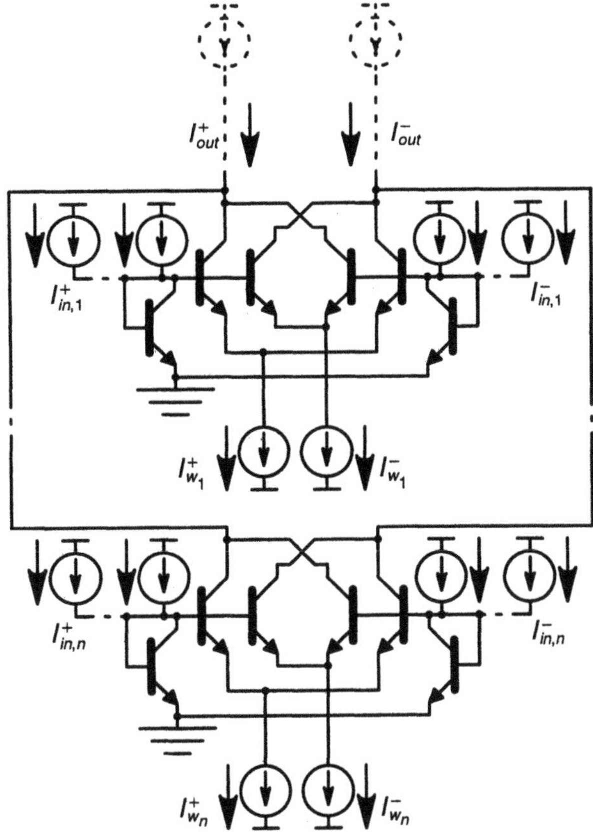

Fig. 5. Summer.

A current normalizing circuit [5] is used to scale the summer tail currents in order to implement (1); see Fig. 6. Currents I_1–I_9 represent "raw," i.e. unscaled, weights. The normalizing circuit produces scaled weight currents I_{w1}–I_{w9}, whose sum always equals I_w, and whose ratios equal the ratios of the "raw" weight currents I_1–I_9. Thus by programming the values of I_1–I_9 the weights w_i of a summer are determined independently of the summer overall gain k_s, while programming the value of I_w determines k_s. The latter can be programmed in the range $-40\,\text{dB}$ – $+40\,\text{dB}$. This arrangement leaves the scaling circuitry out of the signal path of the cell. Details of the digital control of the summer are discussed in Section 4.2.

3.2. *Multiplier*

The multiplier [4] is obtained from the basic circuit in Fig. 4 by replacing the two tail current sources I_B^+ and I_B^- with signal inputs. Instead of the second summer output, a constant can be connected to the second input of the multiplier. The sign of the multiplier output is also programmed.

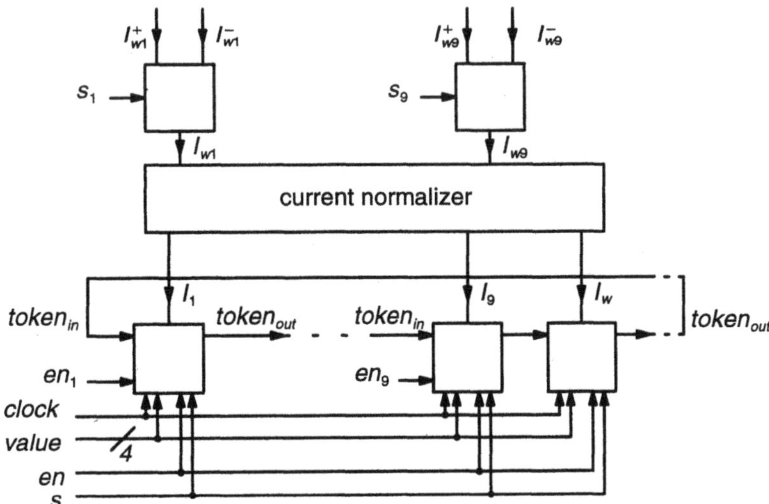

Fig. 6. Controlling the weights of the first summer.

3.3. Amplifier/Integrator

Integration is one of the basic linear signal processing operations, and as such it should be included in a programmable analog device. In many FPAA applications, only some cells will perform integration, therefore the FPAA cell should provide means for turning integration off.

It is easy to implement an amplifier/integrator if some kind of electronic switches, such as MOS pass-transistors, are available. Switches can be used to program the unity-gain frequency (by connecting or disconnecting a number of capacitors), or to turn the integration on and off.

There are at least two problems with switches: (1) they are not easy to implement in some technologies, such as bipolar, (2) they introduce parasitic time-constants which can severely degrade the frequency response of the circuit.

A successful implementation of a switchless current-mode Miller amplifier/integrator in a bipolar transistor-array technology has been demonstrated [16].

The input buffer k_1 (Fig. 7a) comprising transistors Q_{11}–Q_{16} (Fig. 8) is based on a current amplifier of Fig. 4. Only one of the buffer outputs is active at a time, depending on which one of the bias sources I_{E11}, I_{E12} is on.

(a)

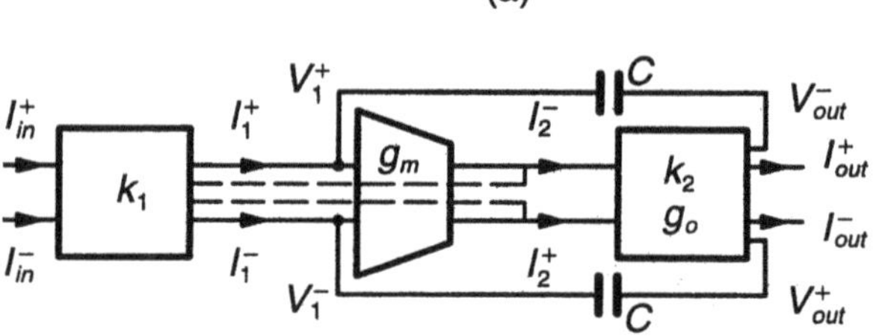

(b)

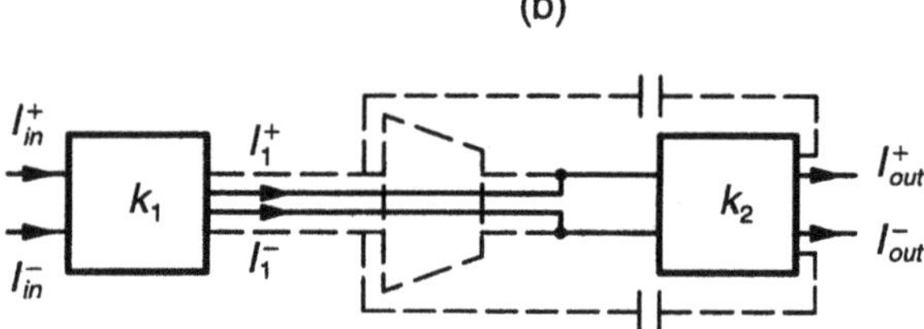

Fig. 7. Current-mode amplifier/integrator: (a) integrator, (b) amplifier.

In the integrating mode (Fig. 7a) sources I_{E12}, I_{C15} and I_{C16} are off. The first output of the buffer which is connected to the simplified g_m cell (Darlington pairs

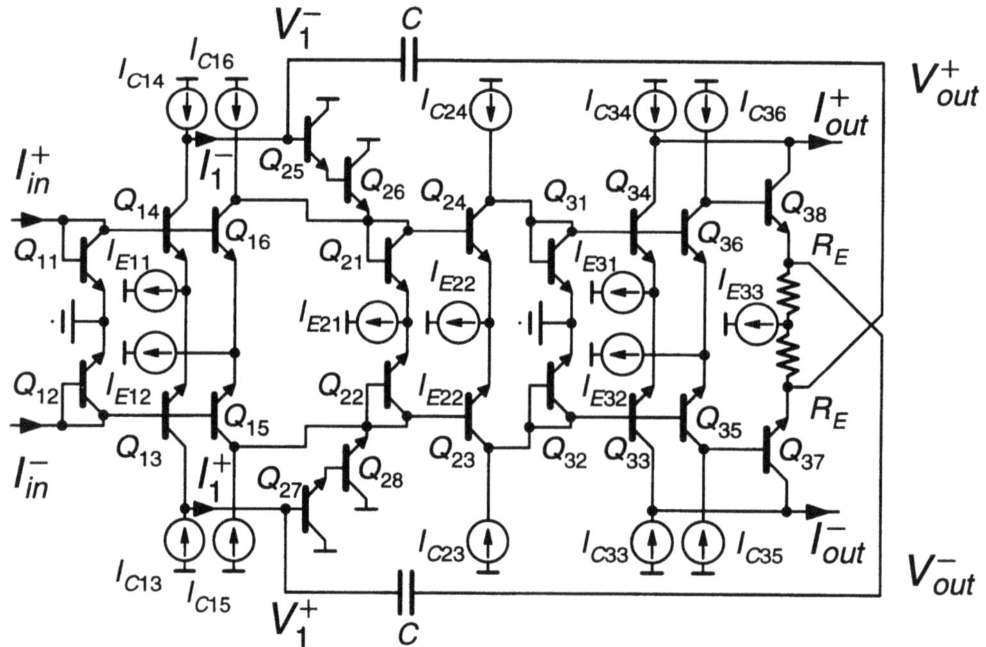

Fig. 8. Simplified schematic of the amplifier/integrator.

Q_{25}–Q_{26}, Q_{27}–Q_{28} and to the capacitors C) is active. A two-stage current amplifier k_2 (Q_{21}–Q_{24}, Q_{31}–Q_{32}, Q_{35}–Q_{36}) follows g_m. Q_{35}, Q_{36} with active loads and emitter follower Q_{37}, Q_{38} provide voltage output. With I_{E31} off, differential output current is I_{C33}, I_{C34} minus collector currents of Q_{37}, Q_{38}. With capacitors C this is a classic Miller integrator in differential form, with an additional current output. The gain (unity-gain frequency) can be changed by changing the bias of the input buffer.

In the amplifying mode (Fig. 7b), I_{E11} is off, the g_m cell receives no signal, and I_{E32}, I_{E33} are off. Buffer k_1 feeds its output current directly to the amplifier k_2 (from collectors of Q_{15}, Q_{16}). The gain of this cascade can be turned up to 60 dB by changing the bias [3,5]. Differential output current is I_{C33}, I_{C34} minus collector currents of Q_{33}, Q_{34}.

Figs. 9 and 10 demonstrate the frequency response in the integrating and amplifying modes, respectively. Adjustment of I_{E33} allows fine tuning of the phase response in the vicinity of $-90°$.

Two common-mode feedback circuits (not shown), assure proper voltage levels at the input of the g_m cell, and the collectors of Q_{35} and Q_{36}. Voltage at the emitters of Q_{21} and Q_{22}, proportional to the common-mode voltage at the g_m input, is compared to a reference level. Correction signals are sent to the bias sources I_{E11}, I_{C13}, I_{C14}. A similar scheme is used for Q_{35} and Q_{36}.

Changing voltage gain within the integrator results in shifting the useful range of frequencies along the frequency axis.

3.4. Clipping Circuits

Each of the clipping blocks shown in Fig. 2 is realized as a circuit of Fig. 4. Additional current sources are connected on the input and the output to enable shifting of the DC transfer characteristic as required. With two blocks one can achieve many nonlinear characteristics, some of them shown in Fig. 3. k_i, z_i, a_i, b_i are the slope, zero, lower saturation level and upper saturation level, respectively (see Fig. 3b).

4. Digital Programming and Control

The characteristics of a particular circuit implemented in the FPAA are determined by the control circuitry, which in general has a twofold purpose:

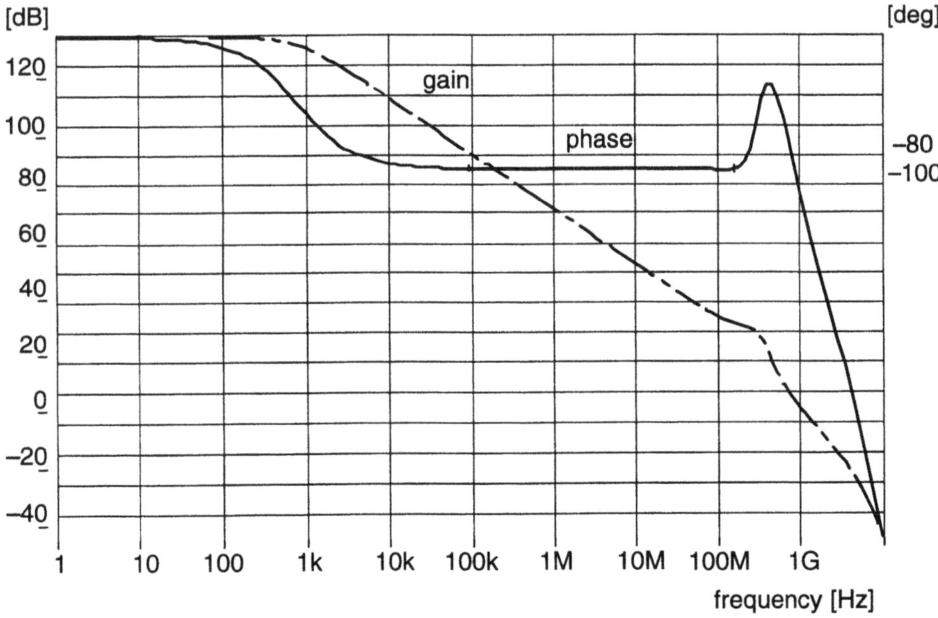

Fig. 9. Integrating mode frequency response.

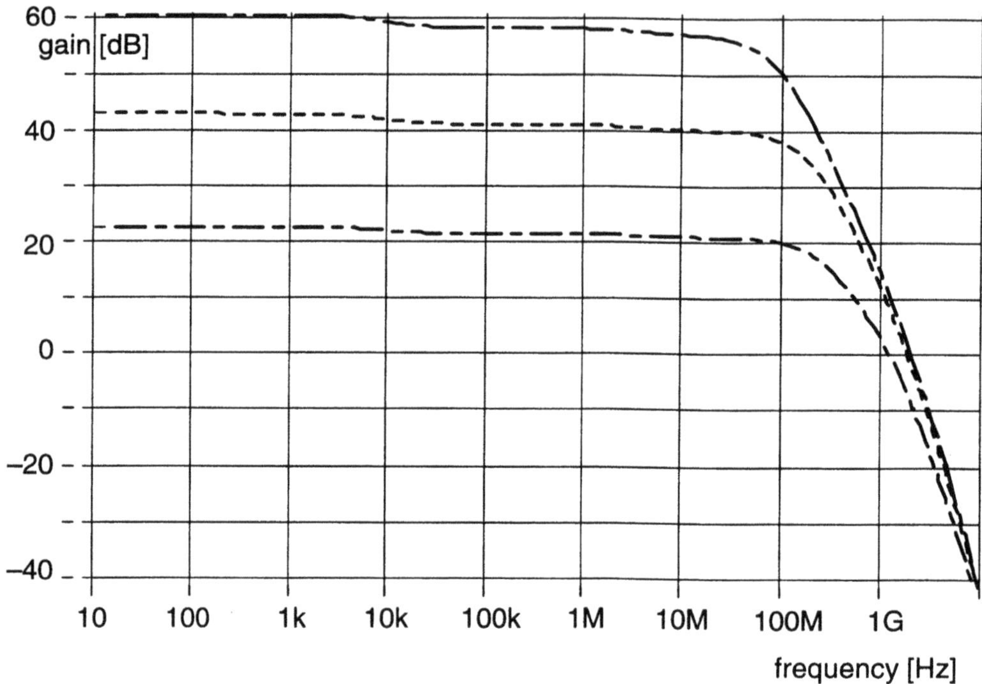

Fig. 10. Amplifying mode frequency response.

1. setting up required functions and parameters of each cell, and
2. realizing the active-control functions of the cell (see Section 2.4.2).

In [15] a general control scheme for these purposes has been proposed.

A modified control scheme, suitable for implementation in a high-density bipolar technology,[3] is presented in this paper. Programming an FPAA requires setting up a number of (i) analog parameters and (ii) binary values, such as the signs of the analog parameters, and the enable bits.

4.1. Parametric Programming

The lack of EEPROM cells and MOS devices in bipolar technologies makes it difficult to design an analog memory cell. A simple analog memory cell and a current source, using JFET devices (available in some bipolar technologies) has been proposed in [20]. The cell holds an analog value for about 200 μs with less than 1% loss, using a capacitor of 0.4 pF. A number of such cells can be connected in a ring, and refreshed using one analog signal line and a single clock signal, and a token passed between the cells. Refreshing 20 such cells would require the clock frequency on the order of 100 kHz.

In high-density bipolar processes[4] it is feasible to use a number of simple digital-to-analog converters (DACs) for the purpose of parametric programming.

The gain of the first summer (i.e. the I_w current (Figs. 2, 6)) is controlled with 12-bit resolution. Each weight current I_1–I_9 is controlled by a 4-bit word. Two more bits: sign s_i and enable en_i are used for each weight w_i. Thus it is possible to set the overall gain of the cell with resolution that is higher than the ratio of any two of the input weights.

The gain k_s of the second summer is constant and equal to 1, and its weights are controlled by 4-bit magnitude words and sign bits. There are no enable bits for the second summer.

The amplifier/integrator's gain (the unity-gain frequency) is programmed by one bit (0 dB or 20 dB).

Each of the clipping blocks parameters is programmed by 3-bit words.

4.2. *The Control Hardware*

All DACs used to control the analog parameters of the cell are connected in a ring (Fig. 11). A *token* bit, passed between the DACs, determines the response of a DAC to its digital inputs. When the token is present (on), the DAC's output current follows the *value* input (Fig. 12). When the token is absent (off), the DAC produces a current corresponding to the last digital *value* latched. The token is "captured" by the DAC on a rising edge of the clock signal when the token input is high. When the token is present, it sets token output to high, enabling the next DAC in the chain to capture the token on the next rising edge of the clock.

4.2.1. Min/Max-Follower. To realize the min/max-follower function (see Section 2.4.2), the en_i (enable) bits of the first summer are controlled by the outputs of the current-mode comparators (Fig. 13). Each of the comparators produces a "1" if the corresponding input signal of the cell (connected to the non-inverting input of the comparator) is greater than the output of the amplifier/integrator, connected to the inverting input of the comparator. When implementing the min-follower function, all comparators whose output is "0" indicate these input signals which are at the moment smaller than the output signal of the cell. These signals are selected on the input of the first summer. The weights w_i of this summer are made equal in order to implement the average of the selected signals. Thus a feedback scheme is formed which results in all the signals presently smaller than the cell output to be averaged to produce the new cell output. In the case of sufficiently slowly changing signals, the output rapidly converges to the true minimum signal, and remains "locked" onto it due to the hysteresis in the comparator characteristic, so long as it is indeed the smallest input signal. Selecting the average rather than the smallest input signal reduces the convergence speed, but makes this feedback scheme much less likely to be "fooled" by a signal that is smallest at the moment only to become larger than other signals a moment later. It also allows simpler hardware to be used.

The *min/max* signal allows inverting of the comparators outputs in order to implement the maximum-follower. The amplifier/integrator works as an amplifier by transmitting the current "maximum" (or "minimum") from the output of the first summer. Its output signal (which is equal to the output signal of the cell) is connected to the inverting inputs of the current comparators. Only the first clipping block is active, with a transfer characteristic shown in Fig. 3a.

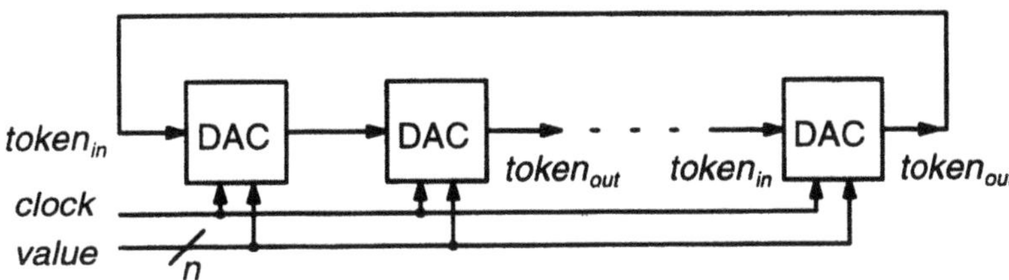

Fig. 11. A ring of DACs.

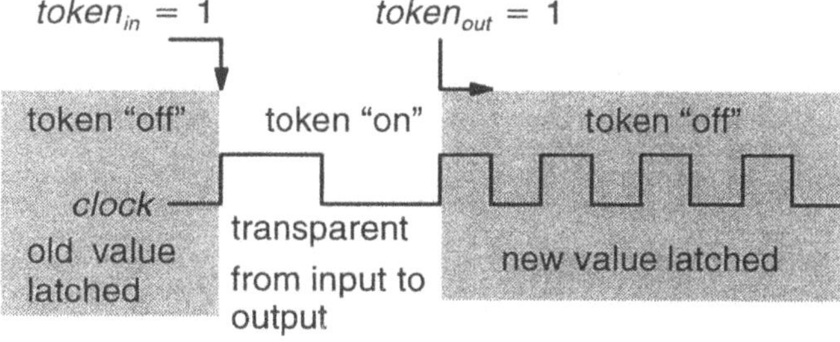

Fig. 12. Token timing diagram.

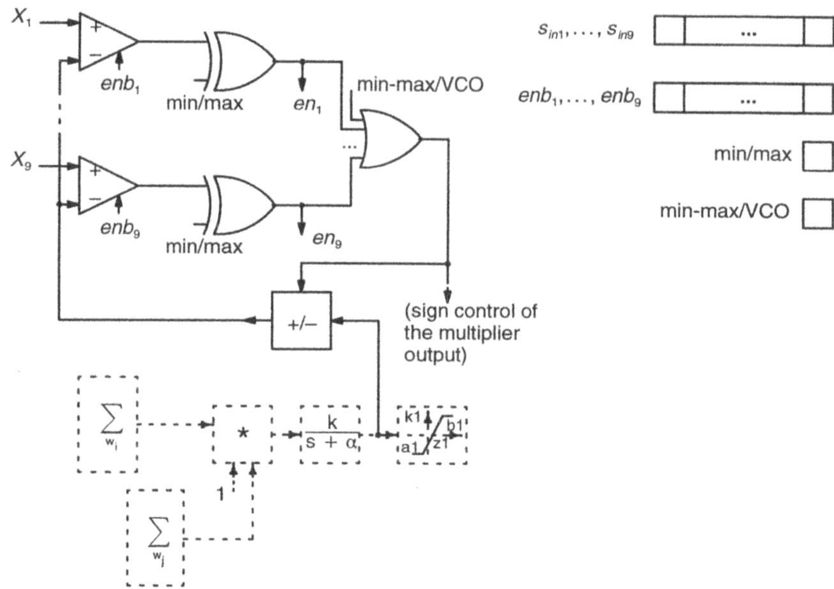

Fig. 13. The control circuitry of the cell.

Fig. 14 shows results of functional simulation of the maximum-follower.

4.2.2. VCO. In the VCO mode, only one comparator in the control block is enabled. The disabled comparators produce a "0" on their outputs. The active comparator's non-inverting input receives a constant signal from the input of the cell. The inverting input receives the output of the integrator, which ramps up or down at the rate determined by the sum of other input signals of the cell. When the output of the integrator achieves the level of the input

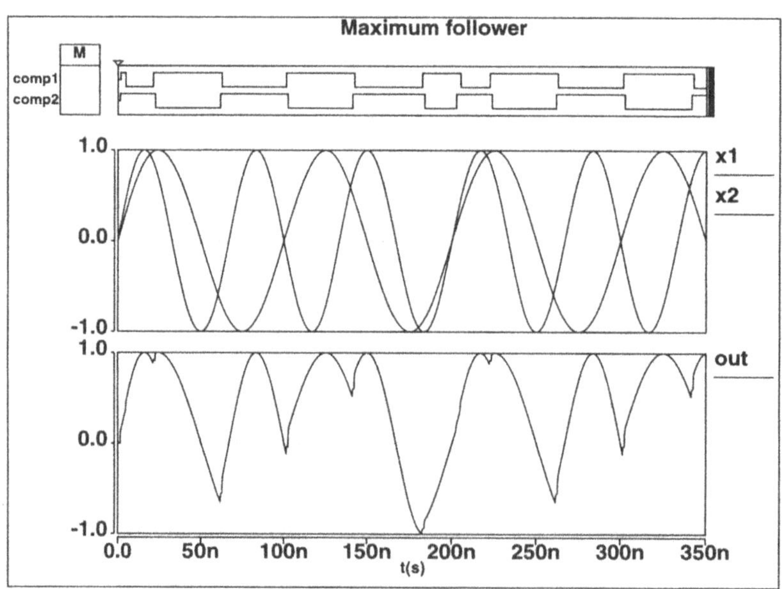

Fig. 14. Maximum-follower operation.

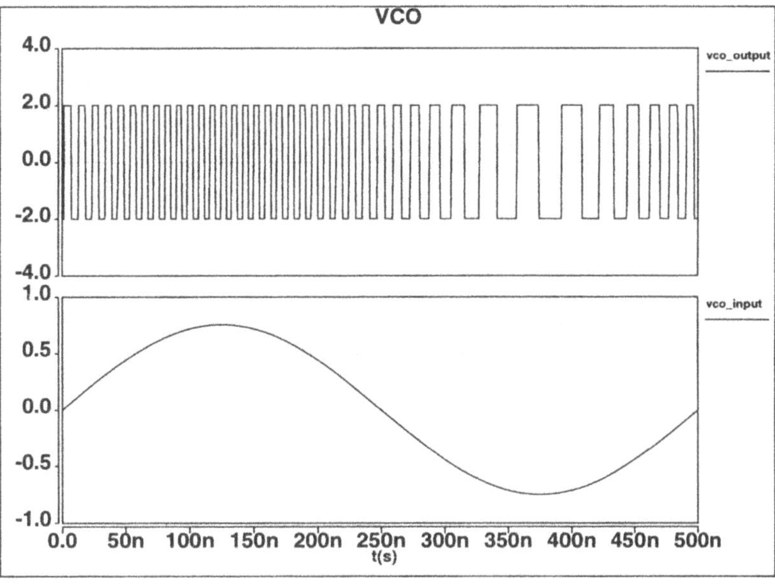

Fig. 15. VCO operation.

threshold signal, the comparator produces (Fig. 13) "1" which propagates through the 9-input OR gate to the analog signal inverter and to the sign bit of the multiplier. Inverting the input of the integrator results in a ramp in the opposite direction. On the other hand, since the integrator output signal passes through the signal inverter, the comparator will again see a signal which ramps up. The process continues to produce waveforms shown in Fig. 15.

The VCO can be controlled by an input signal, or digitally, by changing the relevant input summer weights.

5. Conclusions

The design of a high-frequency, bipolar-technology-based FPAA has been presented. Due to predominantly local signal interconnections and absence of switches in the signal path, high-frequency performance is sacrificed to the smallest possible degree.

A companion paper [17] submitted to this issue demonstrates a variety of applications of the FPAA. These applications effectively prove that limitations imposed on the architecture of the FPAA do not essentially limit its flexibility.

Notes

1. The upper limit on the current gain of a single-stage current amplifier of Fig. 4 is near β of the transistors. When several amplifiers are cascaded, however, controlling of their gain becomes difficult unless the gain is limited to about 10.
2. CPI transistor-array process, Maxim Integrated Products; production-quality models were used for simulation.
3. Such as GST-2 from Maxim Integrated Products.
4. Such as GST-2 from Maxim; up to 200,000 transistors on a die.

References

1. D. R. D'Mello and P. G. Gulak, "Design Approaches to Field Programmable Analog Integrated Circuits." *this issue*.
2. EPAC, "Electronically Programmable Analog Circuit." IMP, Inc., San Jose, Calif.
3. B. Gilbert, "A New Wide-Band Amplifier Technique." *IEEE J. Solid-State Circ.* SC-3(4), pp. 353–365, Dec. 1968.
4. B. Gilbert, "A Precise Four-Quadrant Multiplier with Subnanosecond Response." *IEEE J. Solid-State Circ.* SC-3(4), pp. 365–373, Dec 1968.
5. B. Gilbert, "Current-mode Circuits From a Translinear Viewpoint: A Tutorial." in *Analogue IC Design: the current-mode approach*, ed. C. Toumazou, F. J. Lidgey, and D. G. Haigh, pp. 11–91, Peter Peregrinus Ltd., 1990.
6. F. Goodenough, "Analog Counterparts of FPGAs Ease System Design." *Electronic Design*, pp. 63–73, Oct. 14, 1994.
7. A. B. Grebene, *Bipolar and MOS Analog Integrated Circuit Design*. J. Wiley, 1984.

8. P. R. Grey and R. G. Meyer, *Analysis and Design of Analog Integrated Circuits*. 3rd ed., J. Wiley, 1993.
9. A. Hausner, *Analog and Analog/Hybrid Computer Programming*. Prentice-Hall, Inc., Englewood Cliffs, N.J., 1971.
10. E. K. F. Lee and P. G. Gulak, "A CMOS Field-Programmable Analog Array." *IEEE ISSCC Dig. Technical Papers*, San Francisco, Calif., Feb. 1991.
11. E. K. F. Lee and P. G. Gulak, "A CMOS Field-Programmable Analog Array." *IEEE J. Solid-State Circ.* 26(12), pp. 1860–1867, Dec. 1991.
12. E. K. F. Lee and P. G. Gulak, "Field Programmable Analogue Array Based on MOSFET Transconductors." *IEE Electronics Letters* 28(1), pp. 28–29, IEE, Jan. 2 1992.
13. E. K. F. Lee and P. G. Gulak, "MOS Transconductor-Based Field-Programmable Analog Array." *IEEE ISSCC Dig. Technical Papers*, San Francisco, Calif., Feb. 1995.
14. E. Pierzchala and M. A. Perkowski, "High Speed Field Programmable Analog Array Architecture Design." *Proc. FPGA Workshop*, Berkeley, California, Feb 1994.
15. E. Pierzchala and M. A. Perkowski, "A Field-Programmable Analog Array for Continuous, Fuzzy, and Multi-Valued Logic Applications." *Proc. IEEE ISMVL*, Boston, Mass., May 1994.
16. E. Pierzchala, M. A. Perkowski, P. Van Halen, and R. Schaumann, "Current-Mode Amplifier/Integrator for a Field-Programmable Analog Array." *IEEE ISSCC Dig. Technical Papers*, San Francisco, Calif., Feb. 1995.
17. E. Pierzchala and M. A. Perkowski, "A High-Frequency Field-Programmable Analog Array (FPAA)—Part 2: Applications." *this issue*.
18. R. Schaumann, M. S. Ghausi, and K. R. Laker, *Design of Analog Filters*. Prentice Hall, Englewood Cliffs, NJ, 1990.
19. R. Schaumann, personal communication.
20. O. K. Shana'a, "Circuit Implementation of a High-Speed Continuous-Time Current-Mode Field Programmable Analog Array (FPAA)." MSc thesis, Portland State Univ., 1996.
21. Martin W. Snelgrove, Panel Discussion "On the Future of Analog Circuits", *IEEE ISCAS'96*, Atlanta, Georgia, May 1996.
22. M. A. Tan, "Design and Automatic Tuning of Fully Integrated, Transconductance-Grounded Capacitor Filters." Ph.D. Thesis, Univ. of Minnesota, 1988.

Marek A. Perkowski received his M.S. and Ph.D. degrees from Warsaw University of Technology, Warsaw, Poland. He studied pure mathematics at the University of Warsaw and artificial intelligence in Polish Academy of Sciences. He has been on the faculty at the Institute of Automatic Control, Warsaw University of Technology; Department of Electrical Engineering, University of Minnesota; and is currently a Professor at the Department of Electrical Engineering, Portland State University. His interests are in design automation, logic synthesis, machine learning and digital and analog field-programmable gate arrays. He spent the summer of 1994 in Wright Laboratories, Wright-Patterson Air Force Base, working on application of boolean decomposition to machine learning and was a Visiting Professor at the university of Montpellier and Technical University of Eindhoven in 1996.

He has consulted for several companies in these areas, and also worked for Cypress Semiconductor Corp. as a programmer and system designer of WARP, the first VHDL compiler for EPLDs.

Edmund Pierzchala received his M.S. degree in electronic engineering from Warsaw University of Technology, Warsaw, Poland. He worked as a research assistant and a senior research assistant in the Institute of Biocybernetics and Biomedical Engineering of Polish Academy of Sciences in Warsaw, Poland, and the Nuclear Research Institute in Świerk, Poland. He is presently completing his Ph.D. degree at the Department of Electrical Engineering of Portland State University, where he also taught a number of undergraduate and graduate courses in EE. His research interests include programmable analog circuits, design automation, analog and mixed-signal circuits design, modeling, and simulation.

Analog Integrated Circuits and Signal Processing, 17, 157–169 (1998)

A High-Frequency Field-Programmable Analog Array (FPAA) Part 2: Applications

EDMUND PIERZCHALA AND MAREK A. PERKOWSKI
Department of Electrical Engineering, Portland State University, Portland, OR 97207-0751
edmundp@ee.pdx.edu, mperkows@ee.pdx.edu

Received August 2, 1996; Accepted November 10, 1997

Abstract. This paper presents a variety of applications of an FPAA based on a regular pattern of signal-processing cells and primarily local signal interconnections. Despite the limitations introduced by local interconnections, the presented architecture accommodates a wide variety of linear and nonlinear circuits found in many signal processing systems. Thus it effectively proves that it is possible to improve the performance of an FPAA by means of constraining the interconnection pattern, without significantly limiting the class of circuits it can implement.

Key Words: programmable circuit, analog signal processing, filter, phase-locked loop (PLL), multi-valued logic, fuzzy logic

1. Introduction

A companion paper [24] presents an FPAA architecture based on primarily local signal interconnections and a simple analog signal processing cell design. The purpose of this paper is to demonstrate that architecture limitations for the sake of high-frequency performance do not substantially limit the flexibility of the FPAA, as measured by the number of different classes of circuits that can be implemented in it.

A wide variety of circuits from different classes have been selected in order to demonstrate that a carefully designed FPAA architecture can accommodate them all. This is not to say that one FPAA circuit would suffice for all such applications. Rather, a family of circuits based on a common architecture would be used. For instance, one design might be used to implement linear filters, adaptive filters, etc., another one to implement matrix operations, artificial neural networks, yet another one for fuzzy logic. The selection of applications in this paper serves to demonstrate how a single architecture can be used across many application domains.

The paper is organized as follows. Section 2 presents linear filters. Section 3 shows examples of nonlinear circuits, such as rank filter and phase-locked loop (PLL). Section 4 presents examples of circuits for matrix operations in real time. Finally, Section 5 shows how FPAA can be used to implement multi-valued and fuzzy logic functions.

2. Linear Filters

Fig. 1 shows an electrical schematic of an eighth-order elliptic band-pass filter realized as an OTA-C[1] ladder. This schematic results from the so-called OTA-C simulation [25] of an RLC prototype based on the design presented in [29]. This is a voltage-mode circuit, since each OTA takes a voltage signal as input, and although it produces a current signal, this current is always turned into voltage, either by the integrating operation of a capacitor, or by another OTA with a feedback connection, which is equivalent to a resistor. This is so because eventually each signal created in this circuit is fed to some OTA (which can accept only voltage-mode signals at the input), or is connected to the output terminals, which also require a voltage-mode signal. This circuit, and other voltage-mode circuits, can be realized in an equivalent current-mode form in the structure of the FPAA discussed here in [24].

The network of the filter seems to exhibit little regularity, or locality of connections. Both can be appreciated best when the graph of connections of the filter is drawn (Fig. 2). Each pair of wires carries one differential signal, and it is represented as a single

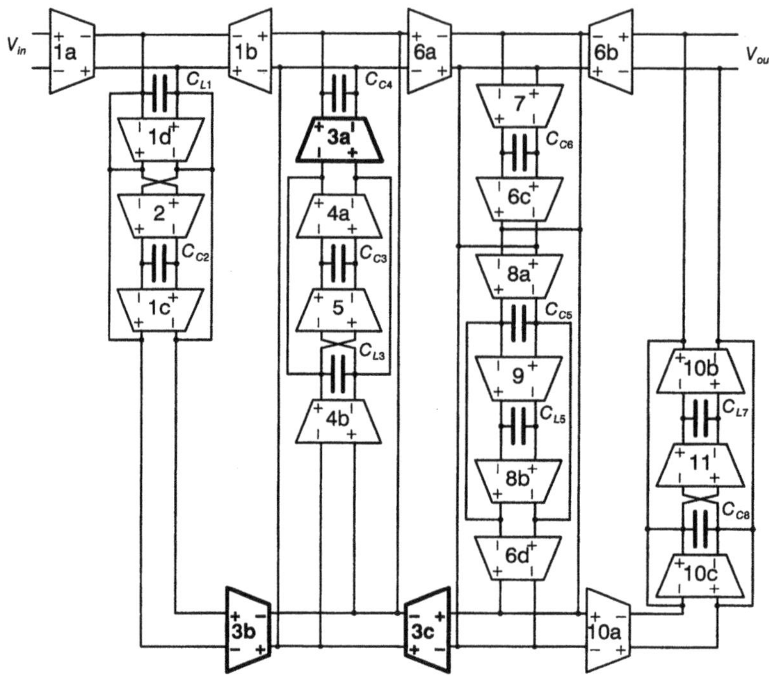

Fig. 1. Eighth-order elliptic band-pass, fully-differential, ladder OTA-C filter.

vertex of the graph. Each OTA is represented as a directed edge of the graph. The graph reveals regularity which leads to a number of different realizations based on regular, locally-only interconnected structures. One way of deriving a regular structure for the circuit is by grouping all edges coming into a given vertex as a single unit. For instance, OTAs 3a-c can be collected together as in Fig. 3a, forming a ''cell'' with four inputs and one output.[2] Eleven such ''cells'' can be connected locally to comprise the entire filter, as shown in Fig. 3b. Dashed lines indicate unused parts of the structure.

Instead of voltage-mode cells (OTA-C), current-mode cells of the FPAA can be used. The pattern of connections is independent of the mode of signals (voltage or current), therefore the same arrangement of cells (Fig. 3b) can be used to map the filter into the FPAA. Each cell works in integrating mode, except cell 6, which realizes ''infinite'' gain (60 dB in the presented FPAA). Cells 1 and 10 realize lossy

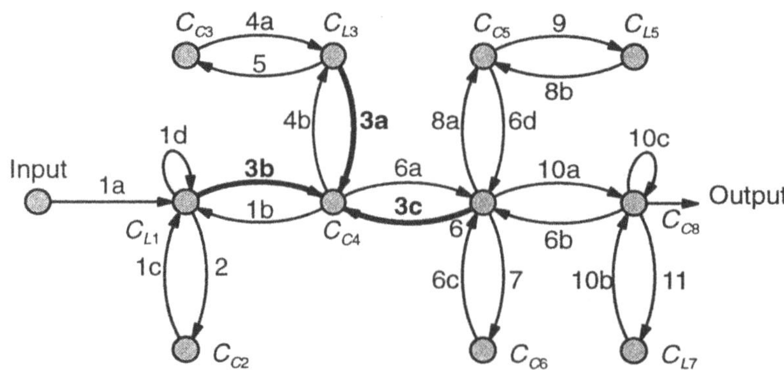

Fig. 2. Graph of connections of the ladder filter.

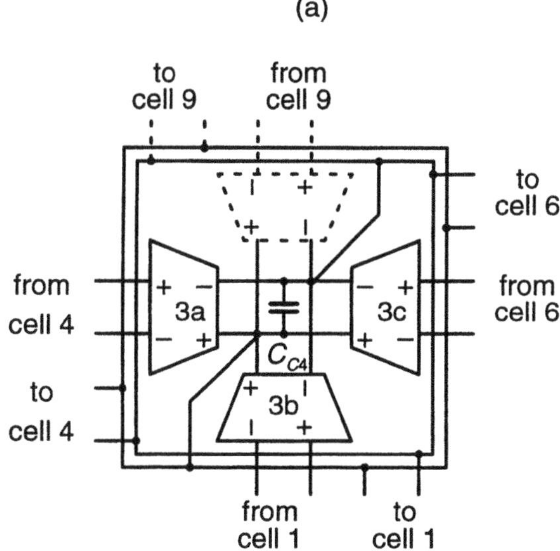

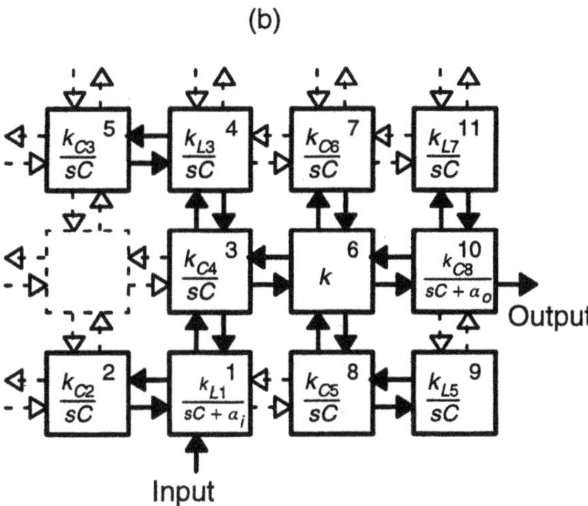

Fig. 3. (a) A single "cell" of the ladder filter. (b) Locally-only interconnected topology of the ladder filter.

integrators, all others—lossless. All the signal interconnections are local, directly between neighbors. Input and output terminals are on the sides of the structure in Fig. 3b, and the circuit can be placed in the corner of the FPAA for easy routing of the input and output. Fig. 4 shows simulated frequency response of the filter.[3]

Most of the ladder filters of practical importance can be mapped into the structure of the presented device in a similar way. Second-order (biquad) filters can be easily mapped, too. Since every transfer function can be realized as a cascade of biquads, which can be then put one next to each other in the array, the device provides a way of realizing continuous-time filters by means of local signal interconnections only.

3. Nonlinear Signal Processing

3.1. Rank Filter

A rank filter [17] can be implemented with only local interconnections. Fig. 5a shows a block diagram of a single cell of the rank filter, and Fig. 5b demonstrates its mapping into the structure of the FPAA. Two FPAA cells are necessary to implement one cell of the rank filter. The left-hand cell in Fig. 5b implements the left-hand part of the rank filter cell, the right-hand cell—the right-hand part. The reader should have no trouble identifying functions performed by each cell in Fig. 5b. A required number of such cells can be easily placed next to each other to realize the rank filter circuit. Fig. 6 shows simulated response of the rank filter.

3.2. PLL

The multiplier in the cell can also work as a phase detector. A VCO can be implemented as shown in [24]. A suitable connection of the two blocks, plus a low-pass filter, implements a PLL (Figs. 7 and 8).

Modulation, demodulation [7], and other functions performed by PLLs can be realized in the FPAA.

3.3. Other Nonlinear Applications

With the multiplier and the clipping circuits it is possible to implement a variety of important nonlinear signal-processing blocks. The multiplier can be directly used as a balanced modulator/demodulator or controlled amplifier. Full wave rectification can be implemented using nonlinear characteristics of the clipping blocks [24]. By combining the latter with low-pass filtration and controlled gain one can implement automatic gain control (AGC).

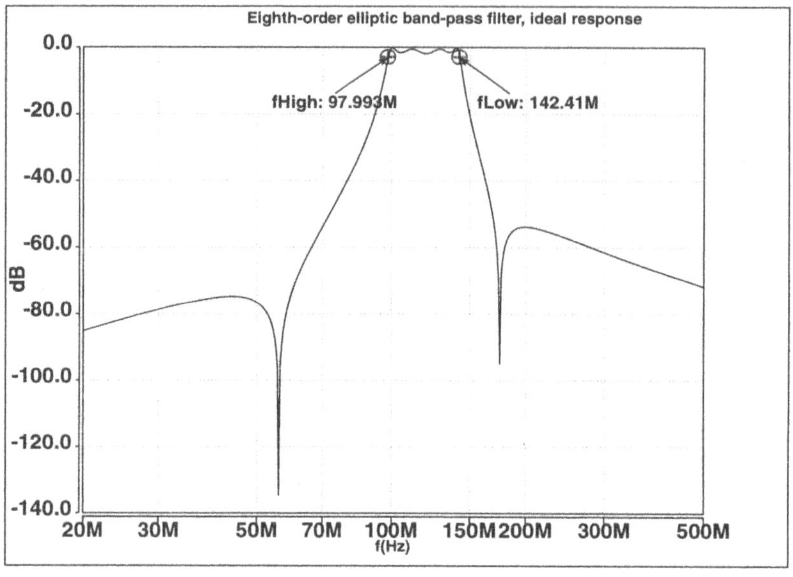

Fig. 4. Frequency response of the ladder filter.

4. Matrix Operations Processor

4.1. Tracking a Matrix Product

Fig. 9 shows the structure of a matrix product tracking circuit. It takes two time-varying matrices $\mathbf{A}(t) = [a_{ij}(t)]$ and $\mathbf{B}(t) = [b_{ij}(t)]$, both 3×3, and creates their product $\mathbf{C}(t) = \mathbf{A}(t) \cdot \mathbf{B}(t)$ (a factor of 3 is required to account for the distribution of each input signal to 3 cells; alternatively the gain of each cell could be increased). Each element $c_{ij}(t)$ of the product matrix is produced by a ''local'' group of cells along a diagonal global signal line. However, to distribute the input signals and to collect the result signals, global connections are necessary. Each diagonal output line is used to sum elementary products $a_{ij}b_{jk}$, $j = 1, \ldots, n$, comprising the product element c_{ik}.

It is instructive to note that the ''globality'' of connections results primarily from the need to distribute input signals, and collect the output signals. Creation of each matrix product is done ''locally'' (although using global signal lines). What is important

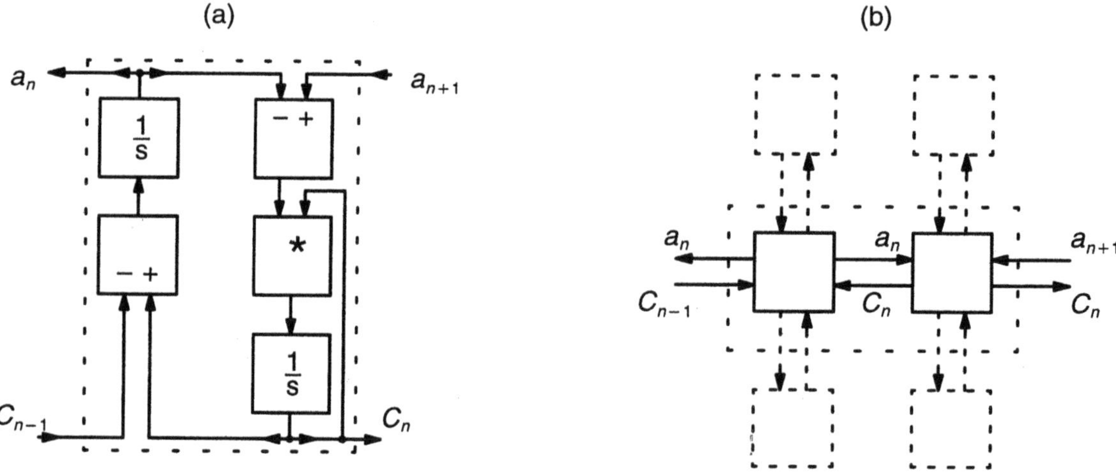

Fig. 5. Analog rank filter cell.

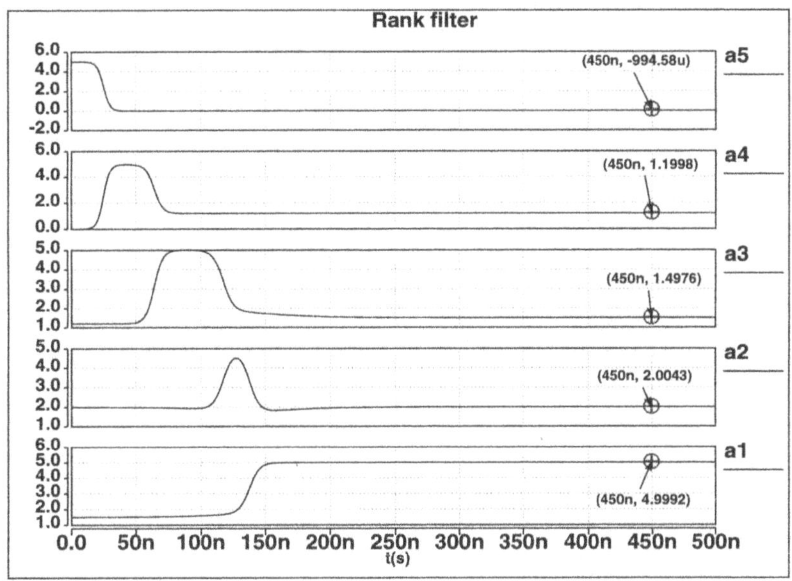

Fig. 6. Analog rank filter operation. Initial conditions: 1.5, 2.0, 1.2, 0.0, 5.0

also, is that global lines are used here only at the "terminals" of the circuit, i.e. for the input and output signals. Global lines are not involved in transmitting internal signals.

The circuit can be easily scaled for any rectangular conformable matrices.

4.2. *Tracking the Solution of a System of Linear Equations*

Modifying slightly the matrix product tracking circuit, one can build a circuit for tracking the solution of a system of linear equations. The solution $\mathbf{x}(t)$ of the system $\mathbf{A}(t) \cdot \mathbf{x}(t) = \mathbf{b}(t)$ can be found by solving a system $\dot{x}(t) + \mathbf{A}(t) \cdot \mathbf{x}(t) - \mathbf{b}(t) = \mathbf{0}$ of differential equations provided that matrix $\mathbf{A}(t)$ is always positive stable [9]. In many practical cases matrix $\mathbf{A}$ will be time-invariant, but it is instructive to see the solution of a more general problem, i.e. with a time-varying matrix $\mathbf{A}(t)$. Fig. 10 shows a circuit solving a system of 3 linear equations with 3 unknowns $x_1(t), \ldots, x_3(t)$. The global connections in this circuit carry internal feedback signals, although the distance traveled by these signals is small.

This circuit can also be scaled easily to accommodate larger systems of equations.

4.3. *Tracking the Solution of a Linear Programming Problem*

A linear programming problem can be stated as follows. Given a set of constraints $\mathbf{g}(t) = \mathbf{F} \cdot \mathbf{x}(t) = [g_1(t), \ldots g_m(t)]' \leq \mathbf{0}$ (the inequality holds for every element of the vector; $\mathbf{F}$ is a rectangular matrix of constraints coefficients, $\mathbf{g}$ is a vector representing individual constraints), minimize the objective function $\rho(x_1, \ldots, x_n) = \rho \cdot x = \rho_1 x_1 + \cdots + \rho_n x_n$, where $\rho = [\rho_1, \ldots, \rho_n]$. Application of the method of steepest descent leads to the system of equations $\dot{\mathbf{x}} = -\mu \cdot \rho' 2a \cdot \mathbf{A} \cdot \mathbf{diag}(\mathbf{g}) \cdot U(\mathbf{g})$ where $U(\mathbf{g})$ denotes the step function, $\mathbf{diag}(\mathbf{g})$ denotes a diagonal matrix with elements of vector $\mathbf{g}$ on the main diagonal, and μ

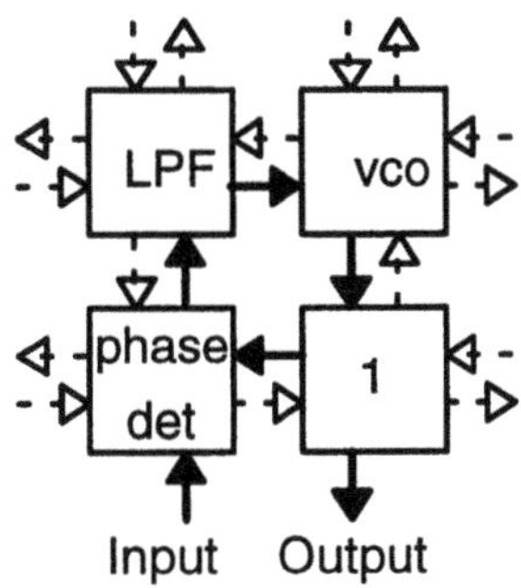

Fig. 7. PLL implementation.

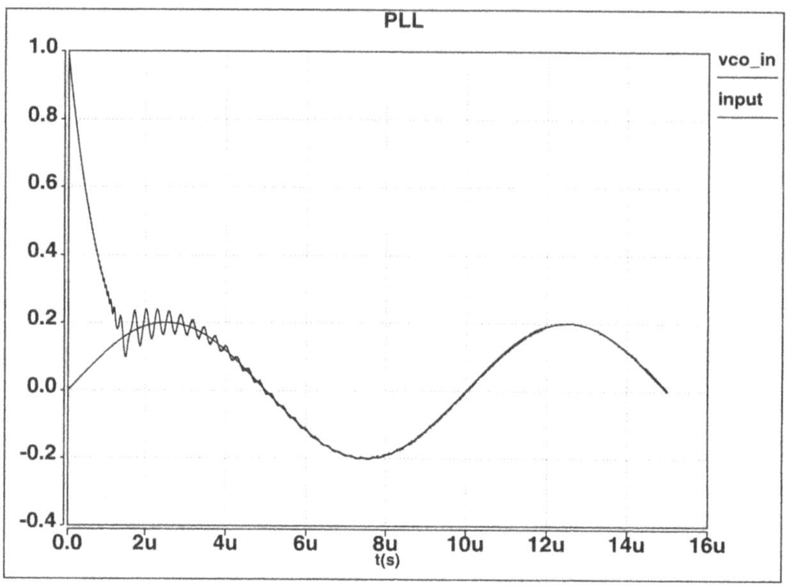

Fig. 8. PLL simulation results.

and a are constants ($\mu \to 0, a \to \infty$), [9]. This system can be solved by the circuit shown in Fig. 11. In the case of linear constraints, matrix **A** is identical to matrix **F**, nevertheless a more general circuit not assuming this equality is shown as an illustration of the versatility of the FPAA architecture. A simplified circuit, with only matrix **F** input, can be easily derived.

Fig. 12 shows results of functional simulation of the circuit.

Like previous examples, this circuit can also be scaled for problems of larger size.

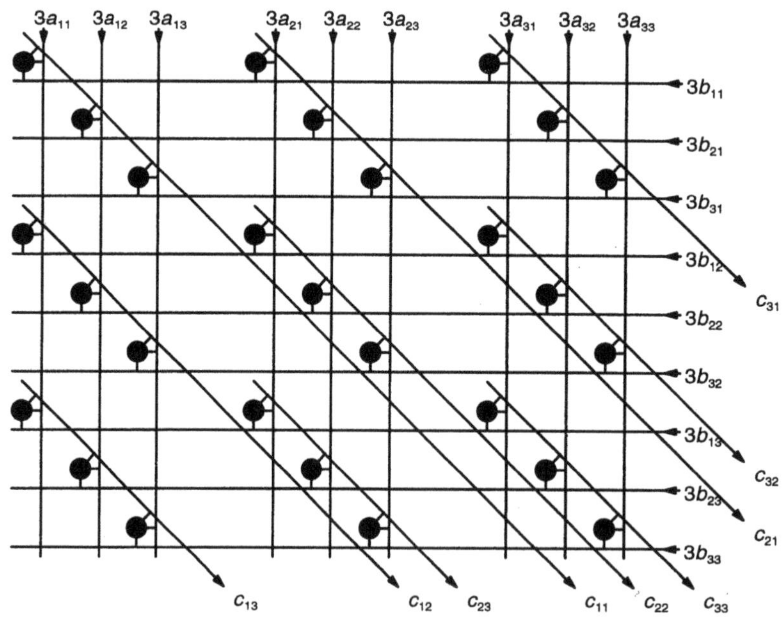

Fig. 9. Continuous-time matrix product tracking circuit.

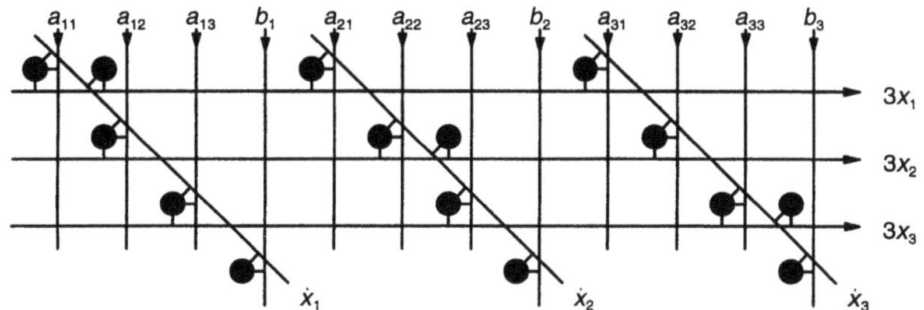

Fig. 10. Continuous-time circuit for tracking the solution of a system of linear algebraic equations.

5. Multi-Valued Logic (MVL) and Fuzzy-Logic Processor

The applications presented in this section require slightly more complex design of the control block of the FPAA cell than that presented in [24]. Fig. 13 shows the enhanced control scheme. The control block can select any of the analog blocks' outputs to be compared with any of the input signals. The selection is made using a summer block, which operates as an analog multiplexer if only one of the weights is non-zero. A programmed constant produced by the control block is available for selection in such comparisons. The control block can adjust any of the analog parameters of the cell. The core of the control block is a simple digital circuit, programmed from the outside of the cell.

Another modification in the control scheme is that the amplifier/integrator can be used as a short-term analog memory (track-and-hold) by cutting off the input signal to the integrator.

5.1. *Galois field 2^2 operations*

Figs. 14a and 14b show the tables for addition and multiplication in a Galois field of four elements. Each of these operations can be realized by a single cell of the FPAA, assuming that only two of the cell's inputs are used at a time. Addition can be realized as $a \oplus b = f(a+b)$ for $a \neq b$ (Figs. 14c and 14d), and $a \oplus b = 0$ otherwise. The condition $a = b$ can be detected by the control block. This requires programming the weights of one of the input summers to calculate the difference $a - b$ of the input signals, selecting constant 0 for comparison in the control block, and controlling the weights of the other input

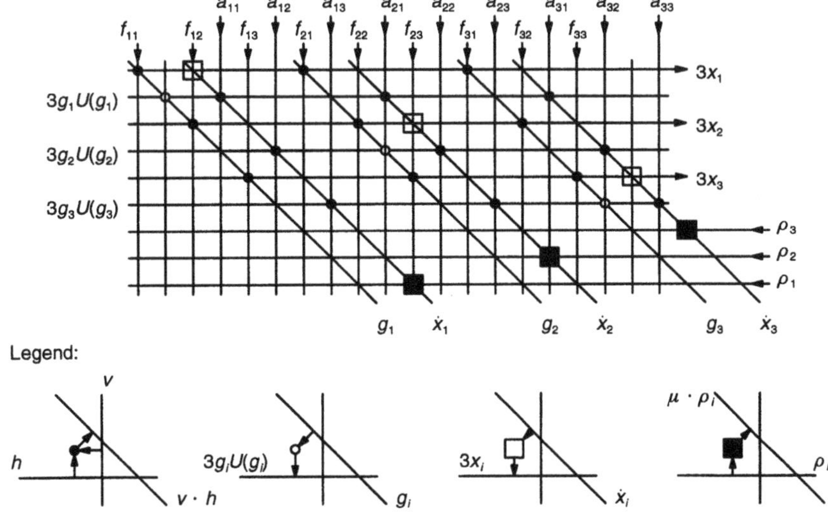

Fig. 11. Continuous-time circuit for tracking the solution of a linear programming problem.

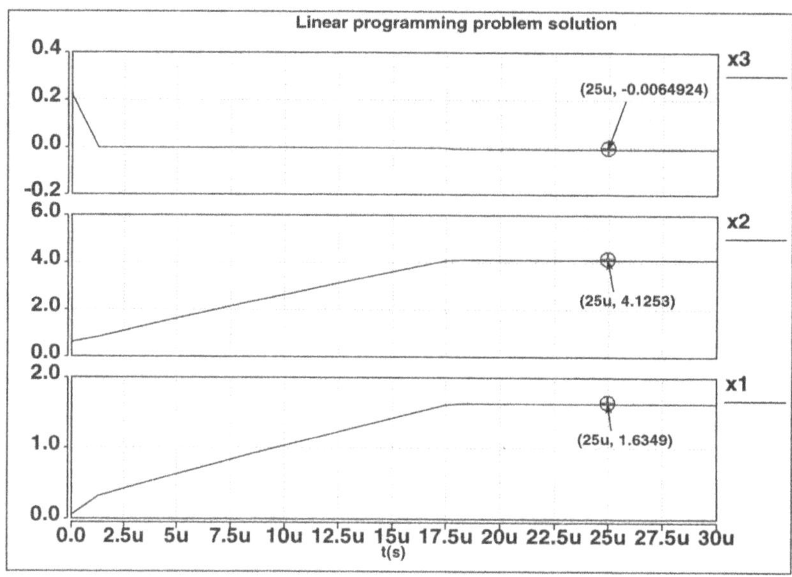

Fig. 12. Linear programming solver operation. Exact solution: (1.6, 4.05, 0.0), [9].

summer to set them to zero if $a = b$ is detected. Instead of function $f(x)$ (Fig. 14d) a smooth function $f_1(x)$ (Fig. 14e) can be used. This function can be realized by adding two characteristics of the clipping blocks shown in Fig. 14f. The function of the form shown in Fig. 14d can be realized by providing more clipping blocks in the cell.

Multiplication $a \odot b$ in the field (Fig. 14b) can be realized as $a \odot b = ((a + b - 2) \textbf{ mod } 3) + 1$ for a $\neq 0$ and $b \neq 0$, and $a \odot b = 0$ otherwise. The two conditions for a and b can be tested independently by the comparators in the control block, and upon at least one of them being true the input weights of the summer would be turned down to 0. The **mod** 3 operation can be realized as shown in Figs. 14g and 14h. The control block performs the necessary logic operations.

The realizations of GF(2^2) operations proposed above are similar to the ones presented in [30].

5.2. *Orthogonal Expansion Structures*

Having defined the addition and multiplication in GF(2^2), we can apply the combinational functions synthesis method based on orthogonal functions proposed in [20]. Fig. 15a shows a block diagram of a structure realizing an arbitrary function of input variables $X_1, X_2, \ldots, X_m$. Each column realizes one orthogonal function over GF(2^2). Multiplied by a constant from GF(2^2), this function is added to the other orthogonal functions. All operations are in GF(2^2). Fig. 15b shows an example realization of one of the functions f_i. More than one column of cells can be used for the realization of any of the f_i's if necessary. Also, it may be convenient to make certain input variables available on more than one horizontal line. An alternative approach, based on providing literals on horizontal lines, or some functions of single variables which are convenient for the creation of literals, is also possible. In one such approach the

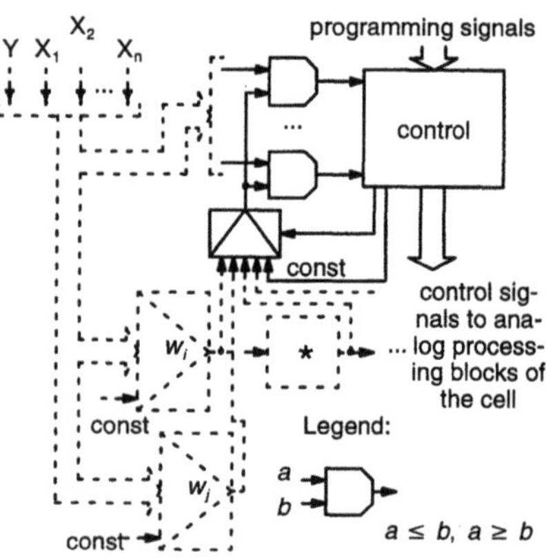

Fig. 13. Enhanced control scheme.

(a)

$a \oplus b$	0	1	2	3
0	0	1	2	3
1	1	**0**	3	2
2	2	3	**0**	1
3	3	2	1	0

(b)

$a \odot b$	0	1	2	3
0	0	0	0	0
1	0	1	2	3
2	0	2	3	1
3	0	3	1	2

(c)

$f(a+b)$	0	1	2	3
0	0	1	2	3
1	1	**2**	3	2
2	2	3	**2**	1
3	3	2	1	0

(d)

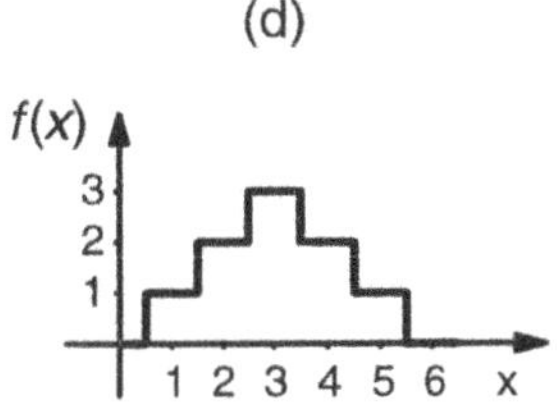

(e)

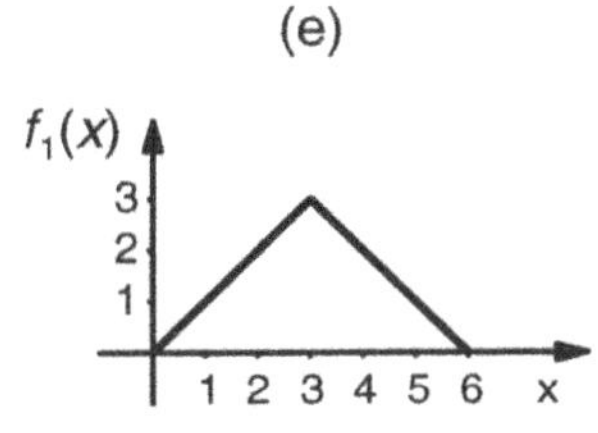

(f)

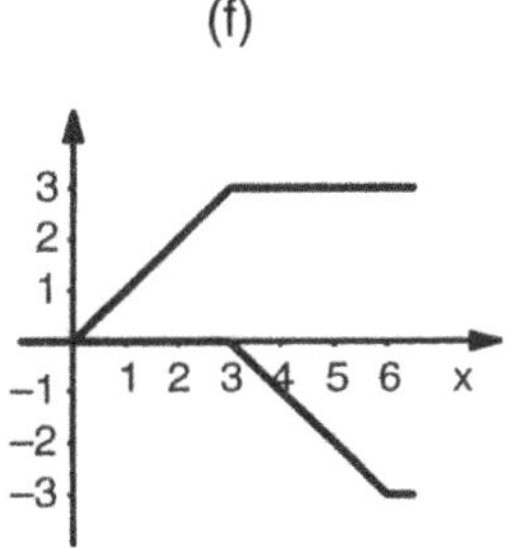

(g)

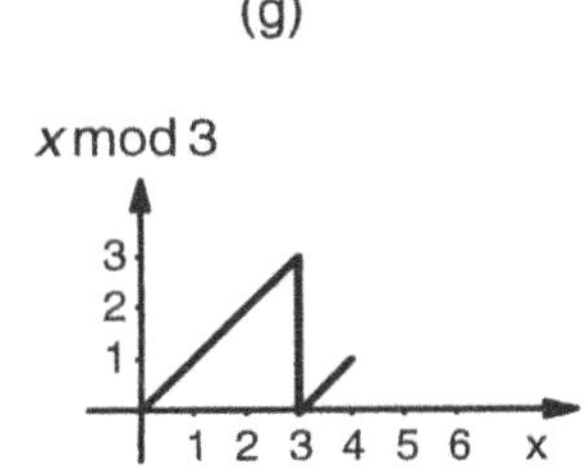

(h)

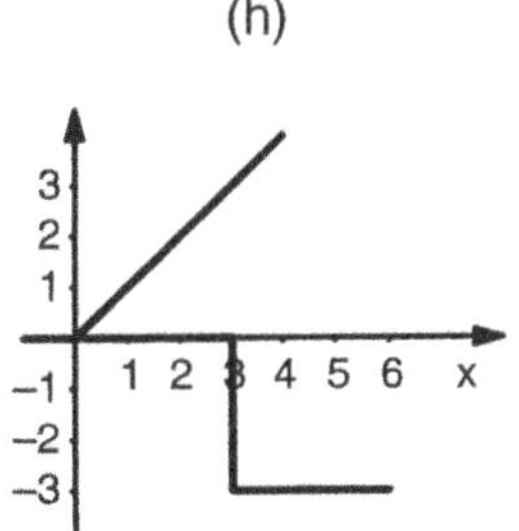

Fig. 14. GF(2^2) operations.

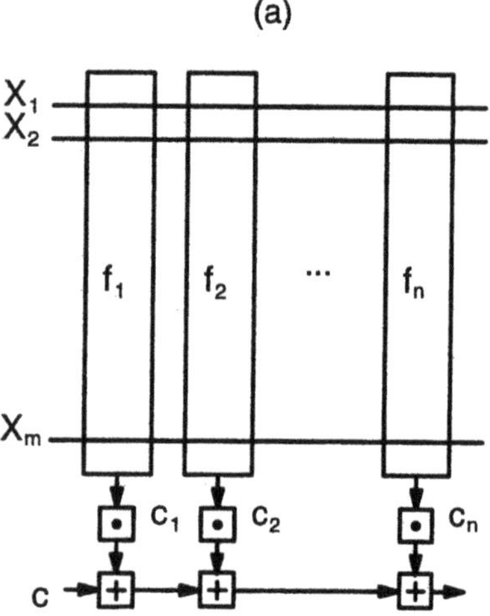

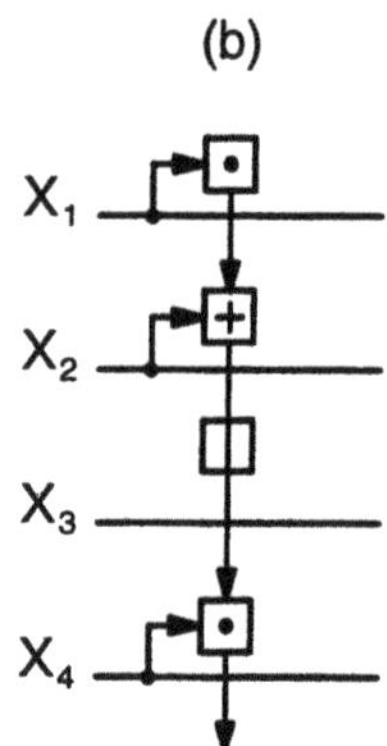

Fig. 15. Orthogonal expansion structures.

powers (i.e. multiple products in GF(2^2)) would be used to create polynomial expansions of mvl functions.

5.3. *Post Logic*

The same structures shown in Fig. 15 can be used for the implementation of Post logic. Each cell can realize **min** and **max** operations [24], and literals of the form shown in [24]. Each function f_i is realized as in Fig. 15b, except that the cells realize **min** or identity operation. Instead of summing over GF(2^2), **max** operation is used.

5.4. *Other Logics*

The structure of Fig. 15a can be used for realization of combinational functions with other methods. Such realizations, unlike the ones based on the orthogonal expansions, may not be unique in the presented structure, however due to the availability of addition, multiplication (in the conventional sense), and non-linear operations on signals, some combinational functions may have very efficient implementations.

Also, the topology of mvl circuits mapped into the FPAA does not have to be constrained to the form shown in Fig. 15. Global vertical and diagonal signal lines can be used, if necessary, to achieve greater flexibility of the circuits topologies. Fig. 16 shows a

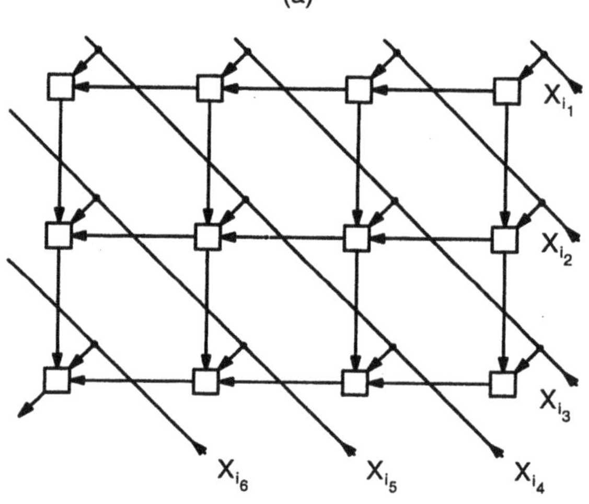

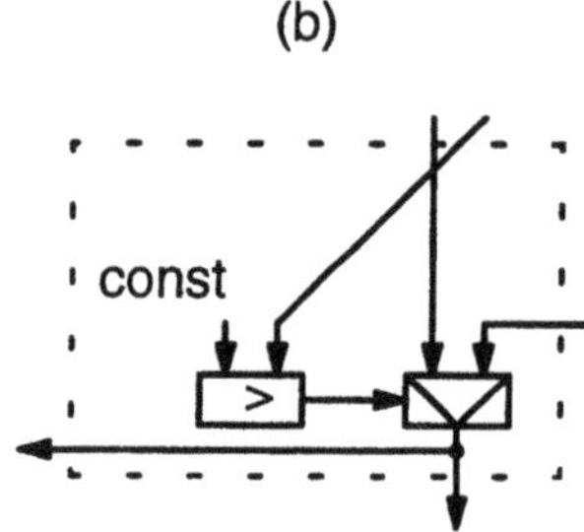

Fig. 16. Generalized Shannon expansion structure.

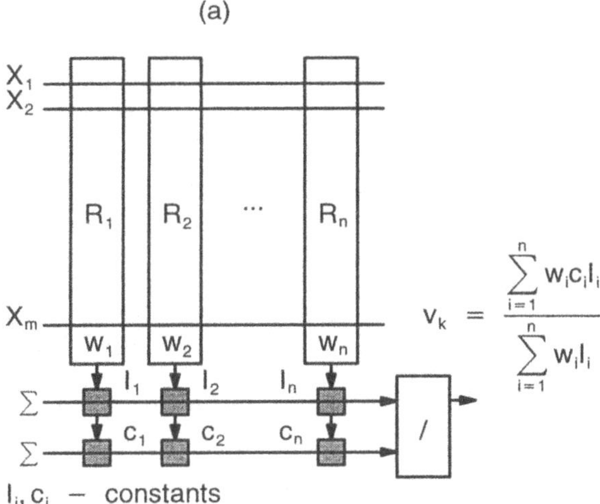

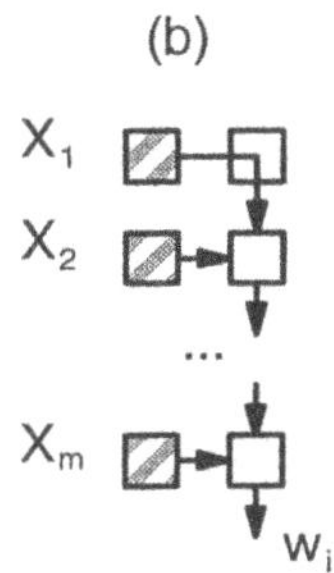

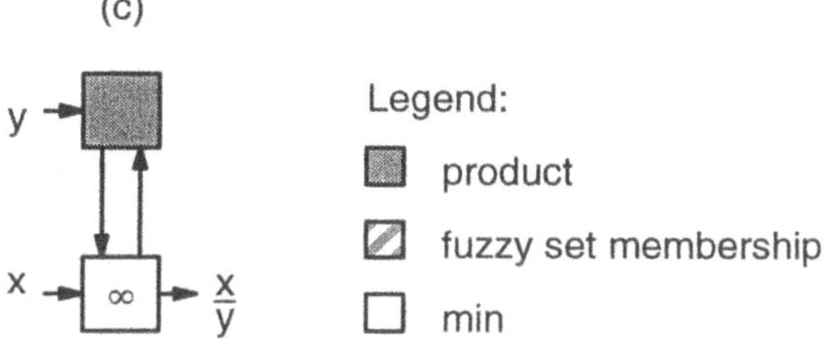

Fig. 17. Fuzzy controller.

structure for implementations based on generalized Shannon expansion of mvl functions [20]. Some input variables need to be connected to more than one diagonal line. More general forms of the same kind are possible, based on operators other than $>$ used for separation, for instance even *vs.* odd parity, based on matrix orthogonality [20], which is a generalization of the approach presented in [18] and [19] for two-valued functions.

Finally, the integrator block can be used as a memory element, enabling realization of sequential circuits. Since each cell can realize the identity function, and global connections are available, larger irregular structures, composed of combinational and sequential parts can be built in the presented FPAA structure.

5.5. *Analog Logics*

Fuzzy logic and continuous logics (such as Lukasiewicz logic) [15,16] can be realized as well. As an example, let us consider an implementation of a fuzzy logic controller with correlation-product inference [10]. A structure very similar to that of Fig. 15a, shown in Fig. 17a, is used to implement a controller with m input variables and n fuzzy inference rules. Fig. 17b shows details of each rule implementation. A fuzzy membership function is implemented as a trapezoidal transfer function of the kind shown in [24]. Activation values w_i are multiplied by centroid values of the fuzzy rules consequents c_i, and their areas I_i, yielding two sums computed on two horizontal global lines. The final expression for the defuzzified output variable v_k is produced by a two-quadrant divider [9] shown in Fig. 17c.

6. Conclusions

A number of circuits from different classes, linear and nonlinear continuous-time, as well as logic, have been shown to map to the structure of the FPAA with predominantly local signal interconnections. This effectively proves that connectivity limitation of the presented architecture does not seriously constrain the range of circuits that can be implemented in the FPAA. Local, and limited global signal interconnections improve high-frequency operation. Continuous-time, switchless design of the analog processing cell allows operation at the speeds near to the limits of a given semiconductor technology.

Notes

1. Operational transconductance amplifier and capacitor.
2. Vertices 1 and 10 additionally require a feedback connection in one of the OTAs to realize lossy integration. All four inputs are used only in cell 6.

3. Functional simulation in Saber (Analogy, Inc.) was used. The cell characteristics were approximated to capture the effects essential for the presented applications.

References

1. EPAC, "Electronically Programmable Analog Circuit." IMP, Inc., San Jose, Calif.
2. B. Gilbert, "A New Wide-Band Amplifier Technique." *IEEE J. Solid-State Circ.* SC-3(4), pp. 353–365, Dec. 1968.
3. B. Gilbert "A Precise Four-Quadrant Multiplier with Subnanosecond Response." *IEEE J. Solid-State Circ.* SC-3(4), pp. 365–373, Dec 1968.
4. B. Gilbert "A Monolithic 16-Channel Analog Array Normalizer." *IEEE J. Solid-State Circ.* SC-19(6), pp. 956–963, 1984.
5. B. Gilbert, "Current-mode Circuits From a Translinear Viewpoint: A Tutorial." in *Analogue IC Design: the current-mode approach*, ed. C. Toumazou, F. J. Lidgey, and D. G. Haigh, pp. 11–91, Peter Peregrinus Ltd., 1990.
6. F. Goodenough, "Analog Counterparts of FPGAs Ease System Design." *Electronic Design*, pp. 63–73, Oct. 14, 1994.
7. A. B. Grebene, *Bipolar and MOS Analog Integrated Circuit Design*. J. Wiley, 1984.
8. P. R. Grey and R. G. Meyer, *Analysis and Design of Analog Integrated Circuits*. 3rd ed., J. Wiley, 1993.
9. A. Hausner, *Analog and Analog/Hybrid Computer Programming*. Prentice-Hall, Inc., Englewood Cliffs, N.J., 1971.
10. B. Kosko, *Neural Networks and Fuzzy Systems, A Dynamical Systems Approach to Machine Intelligence*. Prentice Hall, Englewood Cliffs, NJ, 1992.
11. E. K. F. Lee and P. G. Gulak, "A CMOS Field-Programmable Analog Array." *IEEE ISSCC Dig. Technical Papers* 34, pp. 186–187, Feb. 1991.
12. E. K. F. Lee and P. G. Gulak, "A CMOS Field-Programmable Analog Array." *IEEE J. Solid-State Circ.* 26(12), pp. 1860–1867, Dec. 1991.
13. E. K. F. Lee and P. G. Gulak, "Field Programmable Analogue Array Based on MOSFET Transconductors." *IEE Electronics Letters* 28(1), pp. 28–29, IEE, Jan. 2 1992.
14. E. K. F. Lee and P. G. Gulak, "MOS Transconductor-Based Field-Programmable Analog Array." *IEEE ISSCC Dig. Technical Papers*, San Francisco, Calif., Feb. 1995.
15. J. W. Mills, "Area-Efficient Implication Circuits for Very Dense Lukasiewicz Logic Circuits." *Proc. IEEE ISMVL*, pp. 291–298, May 1992.
16. J. W. Mills, "Lukasiewicz' Insect: The Role of Continuous-Valued Logic in a Mobile Robot's Sensors, Control, and Locomotion." *Proc. IEEE ISMVL*, pp. 258–263.
17. S. Paul, K. Hümper, and J. A. Nossek, "A Simple Analog Rank Filter." *Proc. IEEE ISCAS*, pp. 121–124, San Diego, CA, 1992.
18. M. A. Perkowski, "A Fundamental Theorem for EXOR Circuits." *Proc. IFIP W.G. 10.5 Workshop on Applications of the Reed-Muller Expansion in Circuit Design*, pp. 52–60, Hamburg, Germany, Sep 1993.
19. M. A. Perkowski, A. Sarabi, and F. R. Beyl, "Universal XOR Canonical Forms of Switching Functions." *Proc. IFIP W.G. 10.5 Workshop on Applications of the Reed-Muller Expansion in Circuit Design*, pp. 27–32, Hamburg, Germany, Sep 1993.
20. M. A. Perkowski and E. Pierzchala, "New Canonical Forms for Four-valued Logic." *Internal Report*, Department of Electrical Engineering, Portland State University.
21. E. Pierzchala and M. A. Perkowski, "High Speed Field Programmable Analog Array Architecture Design." *Proc. FPGA Workshop*, Berkeley, California, Feb 1994.
22. E. Pierzchala and M. A. Perkowski, "A Field-Programmable Analog Array for Continuous, Fuzzy, and Multi-Valued Logic Applications." *Proc. IEEE ISMVL*, Boston, Mass., May 1994.
23. E. Pierzchala, "Current-Mode Amplifier/Integrator for a Field-Programmable Analog Array." *IEEE ISSCC Dig. Technical Papers*, San Francisco, Calif., Feb. 1995.
24. E. Pierzchala and M. A. Perkowski, "A High-Frequency Field Programmable Analog Array (FPAA)—Part 1: Design", *this issue*.
25. R. Schaumann, M. S. Ghausi, and K. R. Laker, *Design of Analog Filters*. Prentice Hall, Englewood Cliffs, NJ, 1990.
26. R. Schaumann, personal communication.
27. O. K. Shana'a, "Circuit Implementation of a High-Speed Continuous-Time Current-Mode Field Programmable Analog Array (FPAA)." MSc thesis, Portland State Univ., 1996.
28. Martin W. Snelgrove, Panel Discussion "On the Future of Analog Circuits", *IEEE ISCAS*, Atlanta, Georgia, May 1996.
29. M. A. Tan, "Design and Automatic Tuning of Fully Integrated, Transconductance-Grounded Capacitor Filters." Ph.D. Thesis, Univ. of Minnesota, 1988.
30. Z. Zilic and Z. Vranesic, "Current-mode CMOS Galois Field Circuits." *Proc. IEEE ISMVL'93*, pp. 245–250.

Marek A. Perkowski received his M.S. and Ph.D. degrees from Warsaw University of Technology, Warsaw, Poland. He studied pure mathematics at the University of Warsaw and artificial intelligence in Polish Academy of Sciences. He has been on the faculty at the Institute of Automatic Control, Warsaw University of Technology; Department of Electrical Engineering, University of Minnesota; and is currently a Professor at the Department of Electrical Engineering, Portland State University. His interests are in design automation, logic synthesis, machine learning and digital and analog field-programmable gate arrays. He spent the summer of 1994 in Wright Laboratories, Wright-Patterson Air Force Base, working on application of boolean decomposition to machine learning and was a Visiting Professor at the university of Montpellier and Technical University of Eindhoven in 1996.

He has consulted for several companies in these areas, and also worked for Cypress Semiconductor Corp. as a programmer and system designer of WARP, the first VHDL compiler for EPLDs.

Edmund Pierzchala received his M.S. degree in electronic engineering from Warsaw University of Technology, Warsaw, Poland. He worked as a research assistant and a senior research assistant in the Institute of Biocybernetics and Biomedical Engineering of Polish Academy of Sciences in Warsaw, Poland, and the Nuclear Research Institute in Świerk, Poland. He is presently completing his Ph.D. degree at the Department of Electrical Engineering of Portland State University, where he also taught a number of undergraduate and graduate courses in EE. His research interests include programmable analog circuits, design automation, analog and mixed-signal circuits design, modeling, and simulation.

www.ingramcontent.com/pod-product-compliance
Ingram Content Group UK Ltd.
Pitfield, Milton Keynes, MK11 3LW, UK
UKHW061831190726
13855UKWH00005B/1744
* 9 7 8 1 4 7 5 7 5 2 2 5 0 *